Electroplating Engineering Handbook

Electroplating Engineering Handbook

Edited by Abel Beckwith

www.clanryeinternational.com

Clanrye International,
750 Third Avenue, 9th Floor,
New York, NY 10017, USA

ISBN: 978-1-64726-595-3

Cataloging-in-publication Data

Electroplating engineering handbook / edited by Abel Beckwith.
   p. cm.
Includes bibliographical references and index.
ISBN 978-1-64726-595-3
1. Electroplating. 2. Electrochemistry. 3. Metals--Finishing.
4. Chemical engineering. I. Beckwith, Abel.
TS670 .E44 2023
671.732--dc23

For information on all Clanrye International publications visit our website at www.clanryeinternational.com

Contents

Preface

Electroplating is an electrodeposition process that involves coating a surface with a metal by means of an electric current. Several materials are used for electroplating including copper, nickel, gold, silver and zinc. This process is used to create elements that conduct a larger current such as a micro electrical coil. Electroplating can be applied to deposit relatively thick films of magnetic material to create a larger magnetic force. This mechanism is commonly used in magnetic separation or magnetic force-based actuators. Electroplating can be classified into several types depending on the materials used such as copper plating, silver plating, zinc plating and gold plating. Copper plating is extensively used to prevent case hardening of steel on specified parts. Silver plating is done on tableware, engine bearings and electrical contacts. The most prominent use of gold plating is on jewelry and watch cases. The corrosion of steel articles can be prevented by zinc coatings. The objective of this book is to give a general view of electroplating and its applications. It will provide comprehensive knowledge to all the readers.

This book has been the outcome of endless efforts put in by authors and researchers on various issues and topics within the field. The book is a comprehensive collection of significant researches that are addressed in a variety of chapters. It will surely enhance the knowledge of the field among readers across the globe.

It gives us an immense pleasure to thank our researchers and authors for their efforts to submit their piece of writing before the deadlines. Finally in the end, I would like to thank my family and colleagues who have been a great source of inspiration and support.

Editor

Electrodeposition of Ferromagnetic Nanostructures

Monika Sharma and Bijoy K. Kuanr

Abstract

The fabrication of one-dimensional ferromagnetic nanostructured materials such as nanowires and nanotubes by the electrodeposition technique is discussed. The size, shape and structural properties of nanostructures are analysed by controlling the deposition parameters such as precursors used, deposition potential, pH, etc. The growth of nanostructures and various characterization techniques are studied to support their one-dimensionality. A comparative study of ferromagnetic nanowires and nanotubes is made using angular-dependent ferromagnetic resonance technique.

Keywords: Electrodeposition, nanowires, nanotubes, ferromagnetic resonance, anodic aluminium oxide

1. Introduction

Recently, one-dimensional ferromagnetic nanostructured materials such as nanodots, anti-dots, nanowires and nanotubes have attracted intense research interest. These nanostructures have potential applications in fields as diverse as data storage, magnetics, electronics, optical and microwave devices and nanomedicine [1-6]. They often exhibit new and enhanced properties, due to their low-dimensionality and inter-wire interaction as compared to bulk materials. Several methods have been developed for the synthesis of one-dimensional (1D) nanostructures which lead to well-defined dimensions, morphology, crystal structure, and composition. Various methods such as electrochemistry, chemical reduction, vapour-liquid-solid growth, etc., have been used for preparation of magnetic nanostructures [7-9], although template electrodeposition constitutes one of the most general methods to achieve 1D growth [10-11]. Template-based approach allows to systematically vary the size, shape and structural properties of nanostructures through the modification of template and electrodeposition conditions [12]. The template-based growth often allows the growth of polycrystalline nature of nanowires and nanotubes; however, by controlling the various deposition parameters one can easily overcome such issues and thus *single crystal nanostructures can be achieved*. Electrodeposition technique is also an inexpensive technique which allows the formation of periodically ordered nanostructures in periodic substrates, which enhances their use in potential device applications. The understanding of the growth mechanism would benefit the controlled fabrication of desired metal nanostructures for specific applications. In particular, nanostructures have been prepared in nanoporous membranes by triblock copolymer-assisted hard-template method [13], electroplating [14], a sequential electrochemical synthetic method [15] and nanoporous templates [16]. Fundamentally, nanowire growth depends upon electrodeposition parameters and nanotube formation is dependent on the different growth rates of the metal along the wall surface and from the central bottom of the nanochannels. However, systematic studies are still needed to understand the growth mechanism of nanowires and nanotubes.

Ferromagnetic nanowires and nanotubes exhibit unique and tunable magnetic properties due to their inherent shape anisotropy. Current interest in research on ferromagnetic nanowires and nanotubes is stimulated by their potential applications in different fields such as spintronics, biotechnology, future ultra-high-density magnetic recording media and high-frequency devices [17-21]. For applications such as magnetic recording media, the nanowire's diameter and the inter-wire distance should be as small as possible to increase the areal recording density. Thus, for storage devices magnetic studies concentrate on sub-100 nm nanostructures. The high aspect ratio of the nanowires (Length/Diameter) causes high coercivity, which is helpful in suppressing the onset of the 'superparamagnetic limit', which is considered to be very important for preventing the loss of magnetically recorded information among the nanowires. The inter-wire interaction and magnetic dipole coupling can be controlled by suitable separation among the nanowires.

Several groups have investigated elemental ferromagnetic nanowires such as Fe, Co and Ni with different pore diameters in alumina templates [12-21]. Huang et al. reported single crystal Co nanowires with different diameters and observed that the coercivity and squareness decreases with increase of pore diameter [22]. The size of the crystallites within the Fe, Co and Ni nanowires, as well as the crystalline structure of the nanowires, depends upon the deposition conditions such as the pH of the solutions and the deposition parameters. Giant-magneto-resistance (GMR) properties were found in Co/Cu multi-layered nanowires electrodeposited in nanoporous polymer template [23]. Recently, these ferromagnetic nanowire embedded templates were used as a substrate for making high-frequency devices [24-26].

In this chapter, we shall describe three different aspects of this topic:

1. We first discuss in brief about the template approach for fabricating one-dimensional nanostructures. The most commonly used templates, i.e. anodic alumina membrane (AAM) and polycarbonate (PC) membranes, will be discussed in detail.

2. We will then describe the electrochemical deposition technique used to synthesize ferromagnetic nanowire and nanotube arrays. The size, shape and structural properties of nanostructures are controlled by the type of template used and a number of growth parameters such as precursors used, deposition potential, pH, etc.

3. We will then continue discussing the various characterization techniques and physical properties of these nanostructures. A comparative study of ferromagnetic nanowires and nanotubes has been made using static and dynamic magnetization techniques.

2. Template-based approach

Template electrodeposition is one of the most general techniques to realize one-dimensional growth. In this straightforward approach, the nanostructures are grown electrochemically inside a hard-template material (mould) adopting its shape [27, 28]. Usually, a template (membrane) is a material that has pores of radii varying from a few µm to tens of nm. These porous membranes are generally used in filtration technologies for the separation of different species such as polymers, colloids, molecules, salts, etc. Anodic aluminium oxide (AAO) membrane and polycarbonate (PC) membranes are commonly used membranes for the synthesis of one-dimensional nanostructures [29, 30]. However, PC membranes are disadvantageous as they have random pores and are very flexible. During the heating process, these membranes can lead to distortion, which is one of the major drawbacks for device applications. Also, removal of the template occurs before complete densification of the nanostructures. These factors result in broken and deformed nanostructures. Anodic aluminium oxide (AAO) template overcomes these difficulties as these membranes have uniform and parallel pores and are therefore commonly used for synthesis of 1D nanostructures. Two-step anodization of aluminium sheet in acidic medium solutions of sulphuric, oxalic, or phosphoric acids is used to synthesize AAO templates [29, 31]. Depending upon the anodization conditions, pore densities as high as 10^{11} pores/cm^2 can be obtained and due to mechanical stress at the aluminium-alumina interface, these pores are arranged in a regular hexagonal array as shown in Fig. 1 [32]. A large array of orderly arranged pores with high aspect ratio makes AAO templates ideal for growing nanostructures. The thickness of the template, which determines the length of nanowires, only depends upon the oxidation time and can lead to formation of 1-60 µm thick oxide layers. AAO templates are chemically and thermally inert, causing pure synthesis of ferromagnetic materials. Also, the solution being deposited must wet the internal pore walls and for growth of nanotubules, deposition should start from the pore wall and should proceed inward.

In the anodization process, an electrical circuit is established between a cathode and a film of aluminium which serves as the anode. Then, anodic oxidation or anodization of the film occurs in accordance to the following reaction [33]:

$$2Al + 3H_2O = Al_2O_3 + 3H_2, \ \Delta G° = -864.6 \ kJ \tag{1}$$

where $\Delta G°$ is the standard Gibbs free energy change. During the anodization, initially a planar barrier film forms followed by pore development leading to the formation of the relatively regular porous anodic film, which thickens in time.

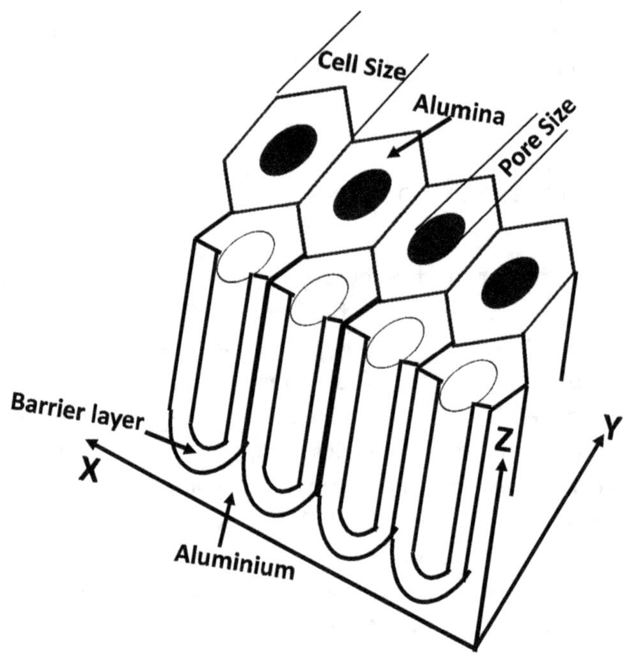

Figure 1. Schematic diagram of the cross section of the porous anodic aluminium oxide showing the nanopores arranged in hexagonal cells.

In two-step anodization, usually the first and second anodization steps could be conducted in the same conditions. The oxide layer formed in the first step is removed by wet chemical dissolution in a mixture of suitable chemicals for an appropriate time, depending on the anodizing time. Based on the applied anodizing voltage and also the type of electrolyte, pores with diameters ranging from few nm to 200 nm can be produced. Li et al. demonstrated a formula between inter-pore distance (D_{nm}) and anodizing voltage (V) [34]:

$$D_{nm} = -1.7 + 2.81\ V\left(volts\right) \tag{2}$$

High-purity (99.99%) aluminium (Al) foil was used for anodization. It was ultrasonically degreased at room temperature in trichloroethylene for 5 min, and then etched in 1.0 M NaOH for 3 min. Before anodization, Al foil was electrochemically polished in a mixed solution of $HClO_4$ and ethanol. A two-step anodization was used to obtain highly ordered pores. Al foil was anodized at 40 V in a 0.3 M oxalic acid at 0 °C for about 12 h in the first anodization step. The oxide layer formed in first anodization was chemically removed in a mixture of phosphoric acid and chromic acid which creates a footprint of nanopores on the Al surface. A second anodization was carried out with the same solutions and steps used in first anodization, resulting in the formation of highly ordered pores [29, 31].

In this chapter, we show results using the commercially purchased AAO templates from Whatman Ltd. and observe that these AAO templates were branched from one side (O-ring support side) as shown in Fig. 2.

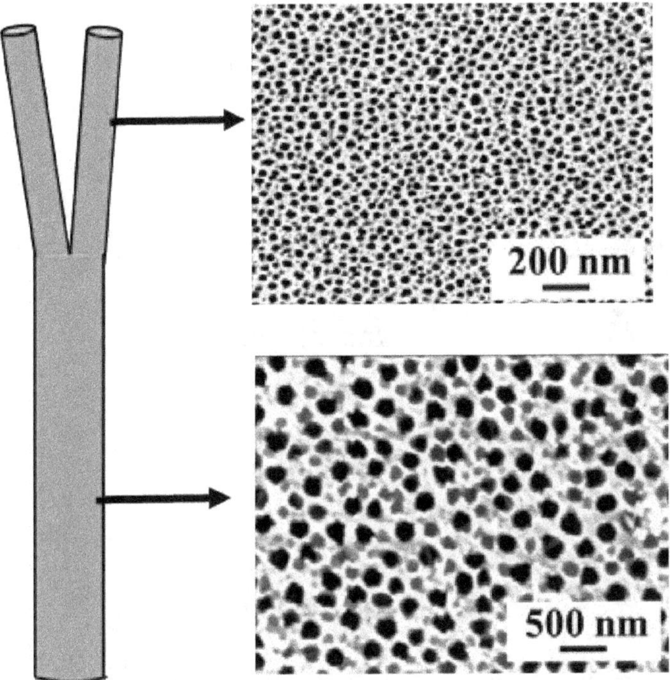

Figure 2. Branching observed in anodic alumina templates commercially available from Whatman.

3. Electrodeposition

Electrodeposition, also known as electrochemical deposition, is a process which involves the oriented diffusion of charged reactive species through a solution when an external electric field is applied, and the reduction of the charged growth species at the deposition surface. Surface charge will develop when a solid is immersed in a polar solvent or an electrolyte solution. The Nernst equation which described the electrode potential is given by

$$V = V_0 + \frac{RT}{n_i F}\ln(a_i)$$

(3)

where V_0 is the potential difference between the electrode and the solution for unity activity a_i of the ions, F is the Faraday's constant, R is the gas constant and T is the temperature.

Electrons will transfer from the electrode to the solution and the electrolyte will be reduced when the electrode potential is higher than the energy level of a vacant molecular orbital in the electrolyte, as shown in Fig. 3. On the other hand, electrolyte oxidation will occur, i.e. the electrons will transfer from the electrolyte to the electrode, if the electrode potential is lower than the energy level of an occupied molecular orbital in the electrolyte, as shown in Fig. 3. When equilibrium is achieved, these reactions will stop. Electrolysis is a process that converts electrical energy to chemical potential in an electrolytic cell where charged species flow from cathode to anode [35].

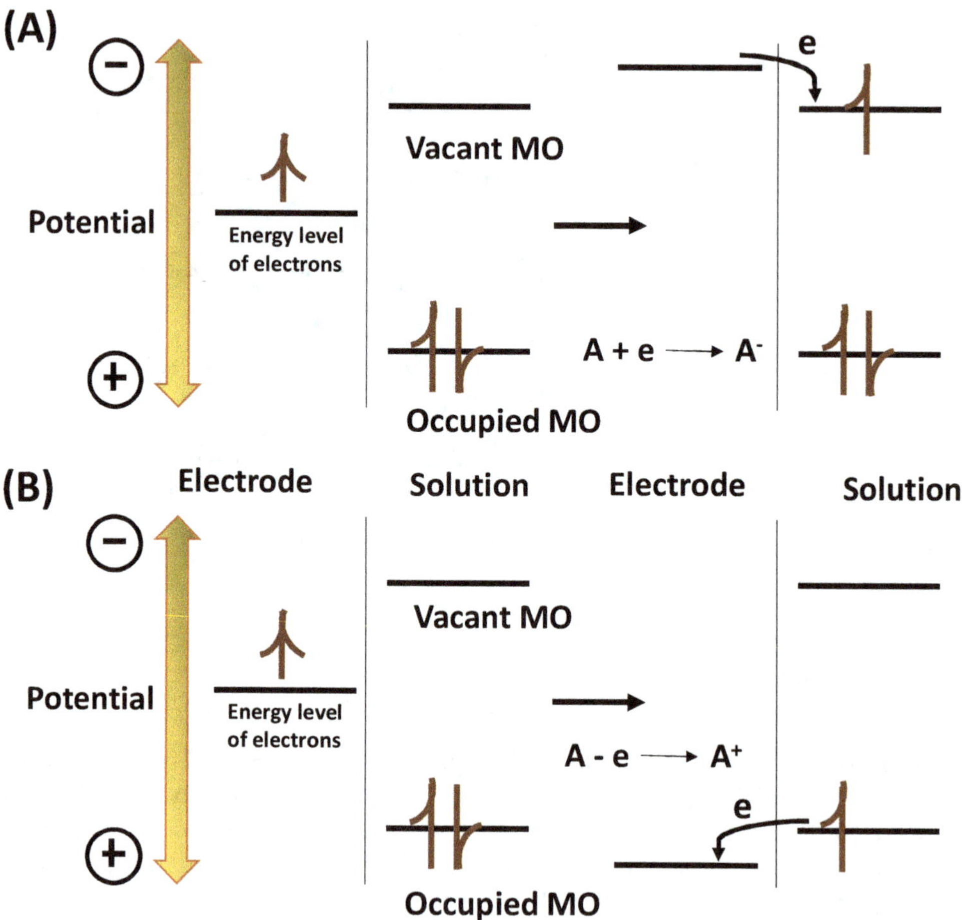

Figure 3. Representation of (A) the reduction and (B) oxidation of a species A in solution. The molecular orbitals (MO) shown for species A are the highest occupied MO and lowest vacant MO [46].

In electrolysis, by selecting the over-potential as means of adjusting the driving force for the reaction, the reaction rate of a system can be easily controlled. With increasing cathodic potential, the concentration of the electrochemical active species, reacting at the cathode, is increasingly reduced in the immediate vicinity of the electrode until every incoming ion is directly reduced. The reaction becomes limited by mass transport processes arising from the depletion of cations in the diffusion layer. The current stays constant even if the over-potential is further increased. This current, limited by diffusion, is denoted diffusion threshold current I_d. In the case of convection, the diffusion layer is of constant thickness δ (stationary case). Mass transport can be described by Fick's first law of diffusion, and the limiting current I_d (at an electrode with the area A) only depends on the bulk electrolyte concentration c_0:

$$I_d = nFAD\frac{c_0}{\delta} \tag{4}$$

The advantages of electrodeposition technique over physical deposition methods are as follows: no need of vacuum equipment, easier handling, higher deposition rates and easier to prepare thick and continuous films. The metal deposition operation depends on a great number of chemical and operational parameters such as local current density, electrolyte concentrations, complexing agents, buffer capacity, pH, levelling agents, brighteners, surfactants, contaminants, temperature, agitation, substrate properties, cleaning procedure. All these parameters act on the structure of the deposit and also on its composition, in terms of alloy and its properties. Accordingly, the determination of these parameters is very important.

Generally, a given metal is electrodeposited from a cell consisting two conductive electrodes, a reference electrode to maintain the potential between the conductive electrodes, an electrolyte and a power supply for conducting an electrical current through the cell or applying an external electric field in the electrolyte. As the current flows into the cell, an oxidation reaction occurs on one electrode (called anode) by charging growth species (typically positively charged metal ions) into the electrolyte and consequently a reduction reaction takes place on the other electrode (called cathode), reducing the charged growth species at the growth or deposition surface as a metallic layer. The process can be carried out using potentiostatic or galvanostatic deposition, depending upon whether applied potential or current is adjusted precisely for the given material to be deposited.

In order to get a qualitative interpretation of the different processes that occur at an electrode, a cyclic voltammogram can be a very helpful tool. The basic idea is that a potential sweep is applied from one to another potential and vice versa with a certain constant speed (mV/s), resulting in a potential triangle (sawtooth) in time (Fig. 4(A)). Because the sign of the potential is changed, oxidative and reductive processes can be distinguished. In addition, varying the scan rate, i.e. triangle slope, different processes at the electrode can be measured. Figure 4 shows the triangle slope and the current flowing through the cathode as function of applied potential.

A similar method has been frequently used for electrodeposition of magnetic nanowires into nanoporous membranes. A metal layer like Au is deposited on one side of the template to provide electrical contact which serves as working electrode, a platinum strip as counter electrode and a saturated calomel electrode (SCE) as a reference electrode. Figure 5(A)

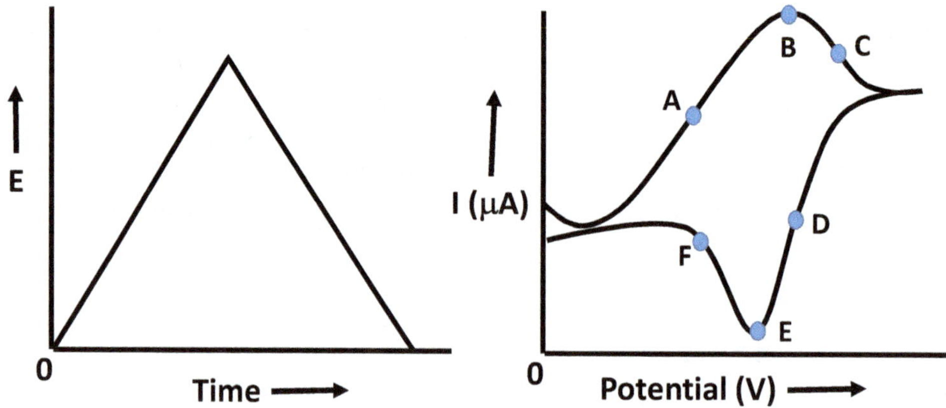

Figure 4. A triangular-shaped potential scan and the resulting current response at a stagnant electrode.

illustrates the common set-up for the template-based growth of nanowires using electrode-position. Chronoamperometry of nanowire deposition is shown in Fig. 5(B). A pre-growth stage occurs when a sudden drop of current takes place once the potential is applied through the cell during which nucleation starts at the pore bottoms. Consequently, a slightly increased current is observed at which the metal is growing in the pores. As pores are filled, current decreases with a large gradient versus time. At the final step, hemispherical caps, originating from each nanowire, form a coherent planar layer that expanded to cover the whole surface of the template. Thus, the effective cathode area increases and a rapidly increasing deposition current can be observed.

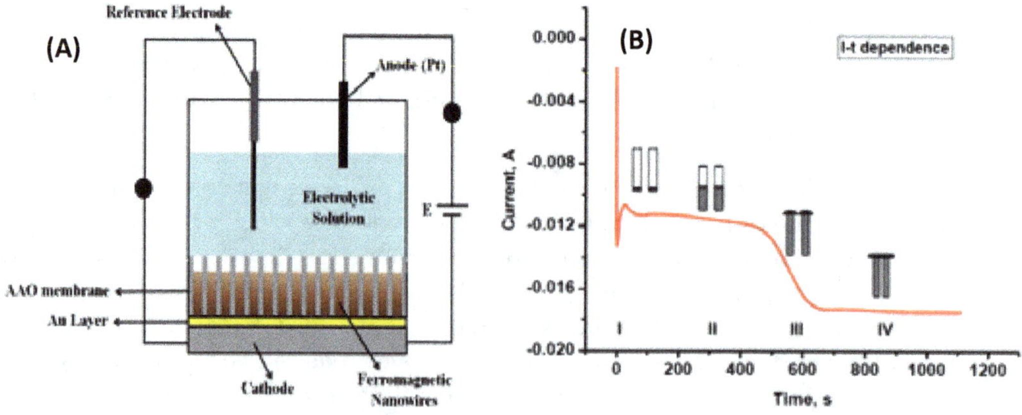

Figure 5. (A) Schematic of electrochemical cell used for nanostructure synthesis. (B) Typlical chronoamperommetry plot during electrodeposition.

For nanotube growth in AAO, instead of a thick Au layer, a very thin layer of Au <30 nm was sputtered on one side of AAO template, so as not to completely block bottom of the pores as

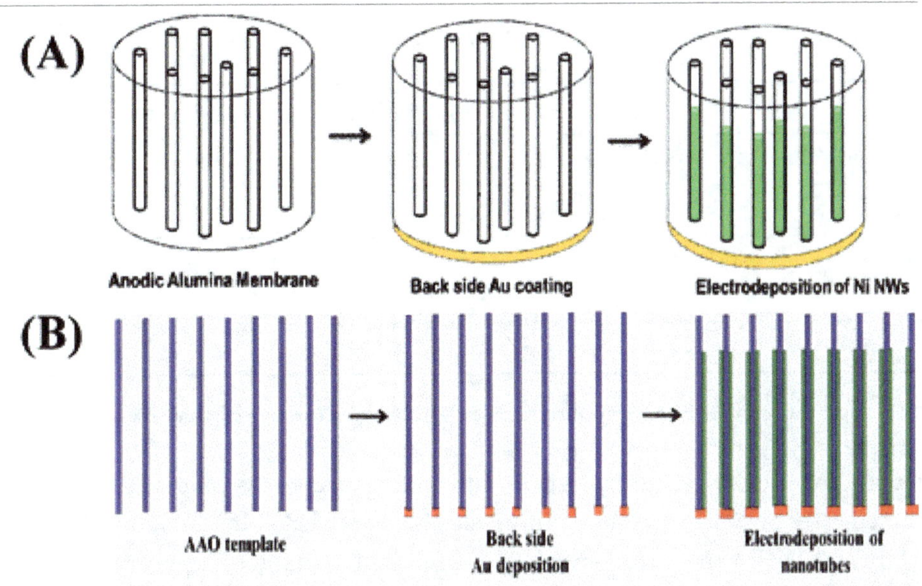

Figure 6. Steps followed for depositing (A) nanowires and (B) nanotubes in AAO template.

shown in Fig. 6(B). Nanotubes (NTs) wall thickness depends upon back-side Au layer as when Ni ion moves to the pores of the template under the drive of electric field, Ni ion firstly reaches the conducting Au layer of the pore-mouth and is deoxidized Ni atom, and thus forms a tubular structure. The wall thickness of nanotubes also depends upon other factors like the current density; therefore, we choose deposition potential in such a manner to achieve low current density. A special attention is paid to address a comparative analysis of the careful preparation of Ni NW (nanowire) and NT (nanotube) arrays with well-controlled ordering and their structural and magnetic response. Table 1 shows the previous research work on ferromagnetic nanowires and nanotubes for various deposition parameters.

Composition	Deposition Potential (V)	pH	Reference
Ni NWs	-1.1 to -1.5	3.5	[36-38]
Co NWs	-1.0	3-5	[38, 39]
$Ni_{80}Fe_{20}$ NWs	-1.2 to -1.4	3	[40]
Ni NTs	-0.2 to -1.5	2-3	[13-15]

Table 1. Electrolytic deposition parameters of various ferromagnetic nanowires and nanotubes.

Electrolytic bath for the deposition of specific ferromagnetic material was prepared using their corresponding sulphate salts; boric acid was added in each bath solution to prevent hydroxide formation and promoting deposition rate. Table 2 describes the electrolytic bath concentrations and electrodeposition parameters used for various nanowires and nanotubes deposited.

Composition	Electrolytes Used	Molar Weight/100 ml	Deposition Potential (V)	pH
Ni NWs	$NiSO_4$, $NiCl_2$, H_3BO_3	33 g, 4.5 g, 3.8 g	-1.0	2.5
Co NWs	$CoSO_4$, H_3BO_3	12 g, 4.5 g	-1.3	3.5
$Ni_{60}Fe_{40}$ NWs	$NiSO_4$, $FeSO_4$, H_3BO_3	12 g, 0.6 g, 4 g	-1.2	3
$Co_{75}Fe_{15}B_{10}$ NWs	$CoSO_4$, $FeSO_4$, H_3BO_3, DMAB	4.95 g, 0.834 g, 4.33 g, 0.049 g	-1.2	2
Ni NTs	$NiSO_4$, PEG, H_3BO_3	1.5 g, 3.7 g, 3.5 g	-0.7	2

Table 2. Electrolytic concentration and deposition parameters of various ferromagnetic nanowires.

The deposition potential of the electrolyte bath with respect to the reference electrode SCE is determined by cyclic voltammetry. Figure 7 shows the typical voltammogram for the Ni, Co, NiFe alloy and CoFeB amorphous nanowires used in this chapter. Typically, the potential is chosen at a reducing potential where current starts decreasing. For the growth of the NWs/NTs chrono-amperommetry is used at various deposition times which controls the length of the nanostructures. The wall thickness of nanotubes also depends upon the current density; therefore, we choose deposition potential in such a manner so as to achieve low current density [37].

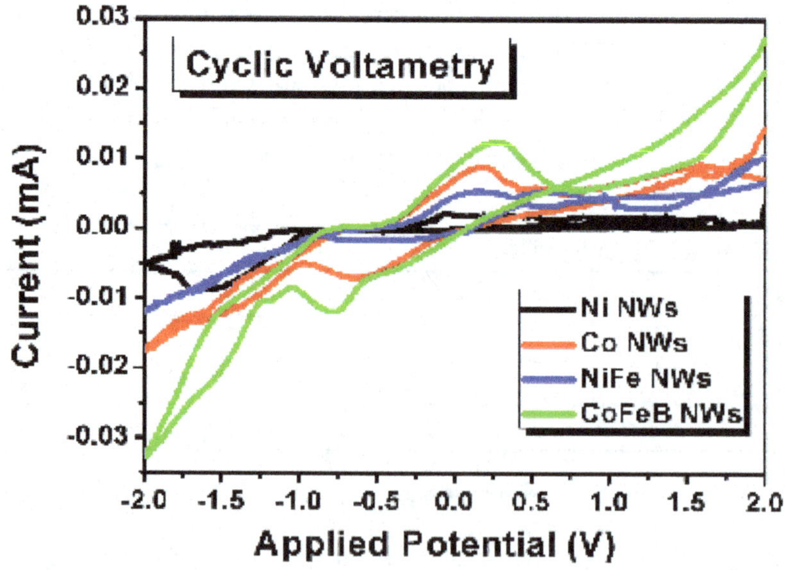

Figure 7. Cyclic voltammogram of various electrolyte baths for determination of deposition potential taken at a sweep of 2 mV/sec.

4. Characterization

In this section, various characterization techniques are discussed to determine the physical and magnetic properties of electrodeposited nanostructures. The morphology and size of nanostructures were characterized by scanning electron microscope (SEM). The qualitative and quantitative elemental composition of nanostructures were done using energy dispersive X-ray analysis (EDX). The structural analysis of ferromagnetic nanowires and nanotubes was done by transmission electron microscope (TEM) and X-ray diffraction (XRD) spectroscopy. Magnetic properties of the samples were tested by superconducting quantum interference device (SQUID).

The morphology of the NWs/NTs was observed by dissolving the template in 3M NaOH solution for 1 h. The separated nanostructures were cleaned several times with de-ionized water and mounted on holder for SEM. Figure 8 (A-D) shows the SEM images of nanowires which revealed that most nanochannels of AAO template are highly filled and parallel to each other. The nanowires have continuous structure without visible defects replicating the pore shapes. It means that ferromagnetic material fills the pores uniformly during electrodeposition under controlled conditions used in this work. Cross-sectional SEM images were taken simply by cleaving of alumina templates. Figure 9 (A-B) shows the SEM images of Ni nanotubes revealing that the wall thickness is around 40 nm and it strongly depends upon the back-side coating of metallic layer.

For EDX analysis, we mounted the nanowire's deposited sample on SEM holder and coated them with carbon since the absorption of X-ray signal by carbon coating is negligible on account of its low atomic number and is taken into account while performing quantitative EDX analysis. Figure 10(A-D) shows the EDX spectrum for various nanowires of Ni, Co, CoFeB and NiFe alloy, confirming the presence of respective elements in the AAO template. The quantitative analysis for CoFeB nanowires determines the stoichiometry of composite material such as $Co_{75}Fe_{15}B_{10}$ and for NiFe alloy such as $Ni_{60}Fe_{40}$.

For TEM characterization, the specimen was obtained by completely dissolving the AAO template in 3M NaOH solution for 24 h and then washing with de-ionized water several times. The free NWs/NTs were then collected from the suspension by applying a magnetic field using a permanent magnet and then rinsed with ethanol and immersed in an ultrasonic bath for 10 min. When operated in the diffraction mode, TEM images also yield information regarding the crystal structure of the nanostructure axis by selected area electron diffraction pattern (SAED).

Figure 11(A-B) shows the TEM image of Co and CoFeB NWs. Figure 12 shows diffraction pattern of the nanowires which revealed the single crystal phase for Co nanowires while the SAED pattern for CoFeB nanowires shows the amorphous phase. TEM images of Ni nanotubes are shown in Fig. 13(A) with inner and outer diameters of about 120 nm and 200 nm. It also confirms that the wall thickness of nanotubes is ~40 nm as observed in the SEM images. The branching due to AAO templates causes the growth of nanotubes in Y-shape nanochannels as observed in Fig. 13(B).

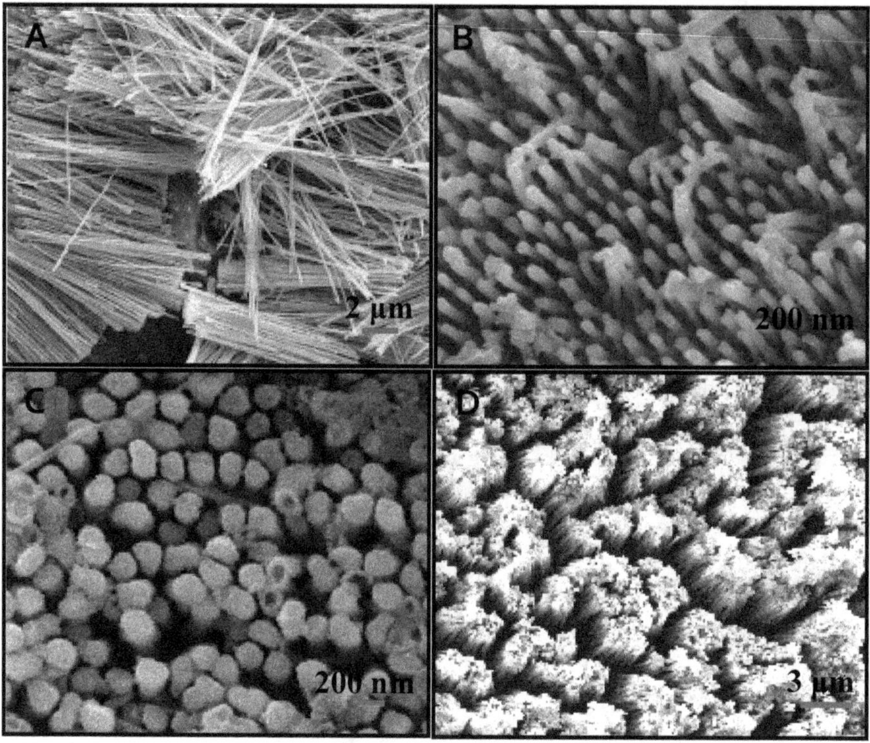

Figure 8. (A-D) shows the top and cross-sectional view of AAO templates before and after the deposition of nanowires. The images confirmed that the diameter of the nanowires are ~200 nm and the interwire spacing between the pores is ~300 nm.

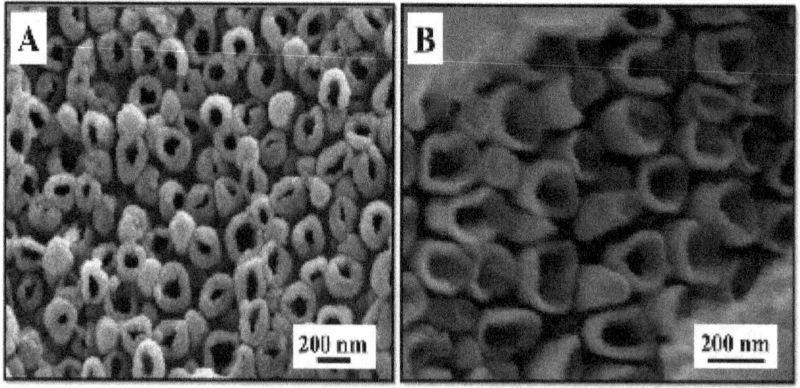

Figure 9. (A-B) SEM images of Ni nanotubes embedded in AAO template with wall thickness of 40 nm.

For determining the crystallographic information of the materials, XRD is one of the most important technique. It can provide information about the crystallographic phases, crystallite size, lattice constant, presence of impurities, etc. X-ray diffraction was performed using Philip's

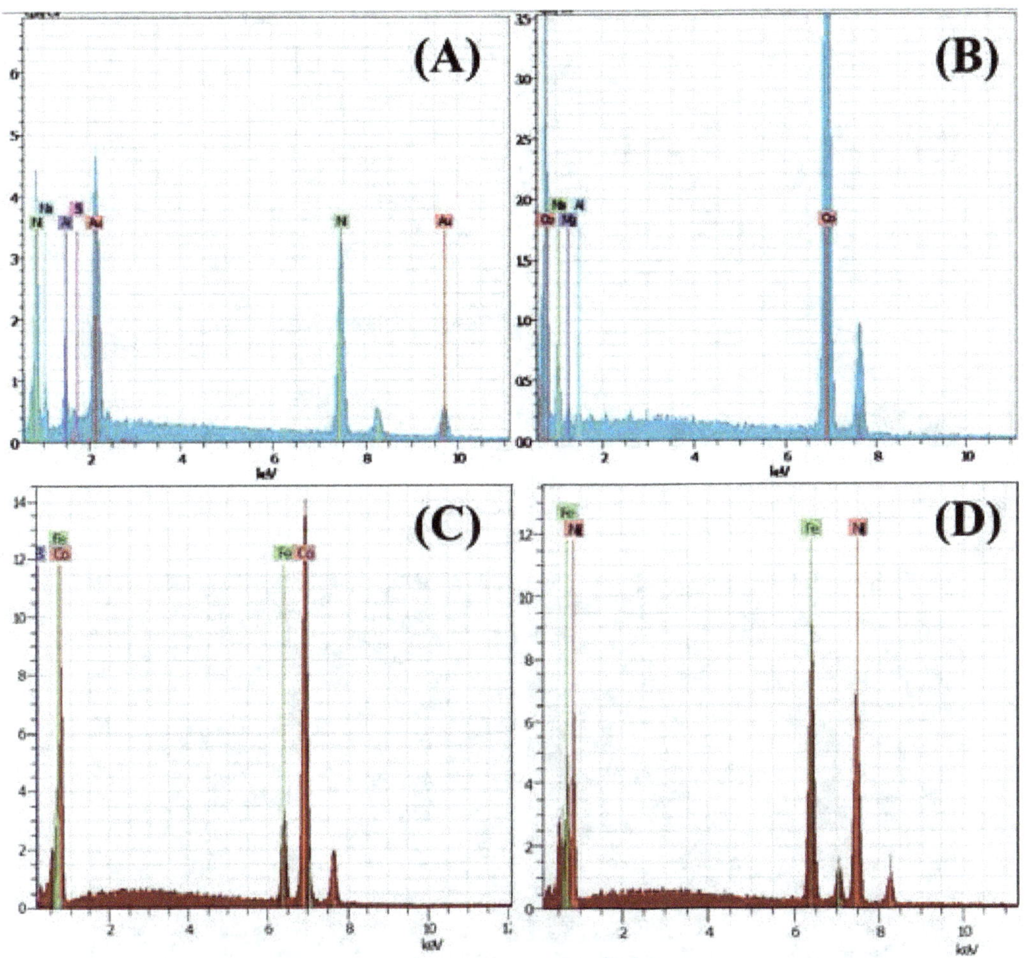

Figure 10. EDX spectra of various nanowires deposited in AAO template (A) for Ni, (B) for Co, (C) for CoFeB, and (D) NiFe alloy.

X'pert PRO (Model PW 3040, X-ray wavelength of Cu Kα line λ = 1.54060 Å) diffractometer. The phase was identified using standard diffraction files of JCPDS. The size of the crystalline material can be estimated from the width of the reflection plane using Scherer equation [41]:

$$D_{crystalline} = \frac{\kappa\lambda}{\beta cos\theta}$$

(5)

where κ is a particle shape factor (for spherical particles, κ =0.9, β is the full width at half maximum (radians) and $D_{crystalline}$ is diameter of the crystallites (Å). It is useful to eliminate the instrumental line width from the observed one to get a correct broadening value due to small particle size.

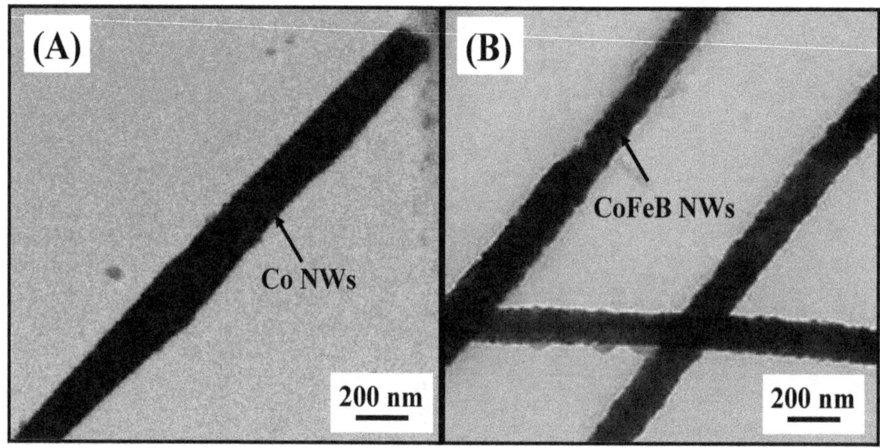

Figure 11. (A-B) TEM morphologies of Co and $Co_{75}Fe_{15}B_{10}$ nanowires with pore diameter of 200 nm.

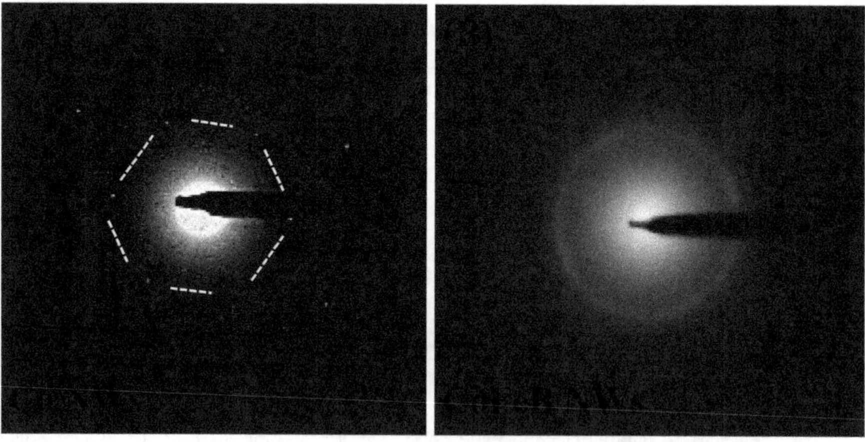

Figure 12. SAED pattern of (A) a single crystal Co nanowire having hcp structure and (B) amorphous phase of $Co_{75}Fe_{15}B_{10}$ nanowires. Dotted lines are guidelines to eye.

Figure 14 shows the XRD pattern of various ferromagnetic nanowires grown in anodic alumina templates by electrodeposition method. The spectra for the nanowires indicates polycrystalline reflection peaks. The diffraction peaks confirms the *fcc* lattice structure in all deposited nanowires except for CoFeB nanowires. The only intense broad peak of CoFeB nanowires is due to a small bcc CoFe (110) phase, which occurs if the content of Fe is less. The diffractogram indicates that CoFeB nanowires appears in the amorphous phase. We compare the XRD spectra of Ni nanowires and nanotubes as shown in Fig. 15. XRD measurements illustrate a face-centred cubic (*fcc*) Ni pattern for NWs and NTs in AAO templates. Compared to the peak positions of standard Ni (JCPDS, 04-0850), peaks are found to be in agreement with peak positions of Ni ((111), (200), (220)).

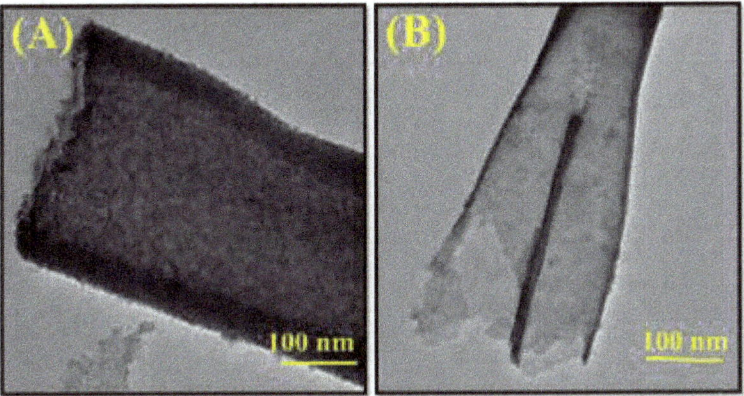

Figure 13. (A) TEM image of isolated Ni nanotube, (B) Y-shape growth of nanotube caused by the branching of AAO templates.

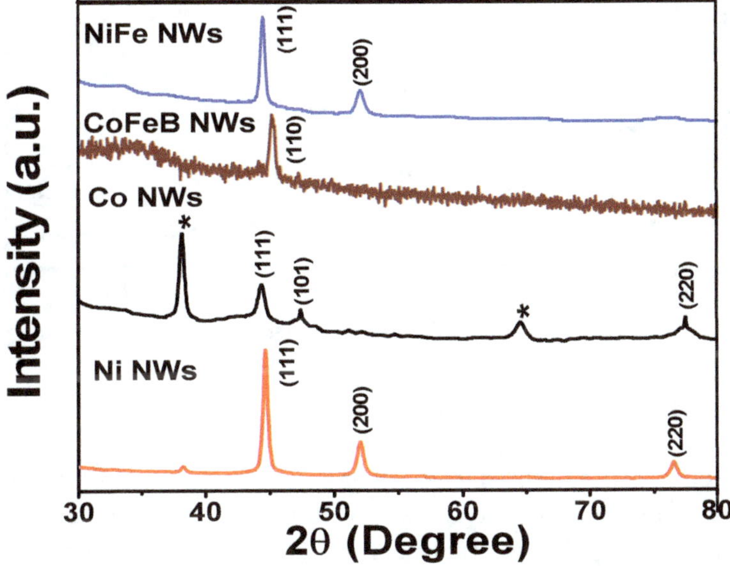

Figure 14. X-ray diffraction pattern of ferromagnetic nanowires electrodeposited in AAO templates. The * peaks in the XRD pattern of Co NWs are due to the Au layer on back side of the template.

The NWs and NTs were found strongly textured along (111) reflection plane. The crystallite size of the deposited nanostructures was determined using Scherrer formula, giving 22 nm and 16 nm for NWs and NTs, respectively. This is due to the fact that their outer diameter is the same, ~200 nm, but NTs have a core cylindrical hole with an estimated diameter of 120 nm. The corresponding d-spacing was observed to be 2.04 nm and 2.03 nm for NWs and NTs, respectively. We also determined the crystallite size of all these nanostructures as given in Table 3.

Material	Peak Position 2θ (Degree)	FWHM β (Degree)	Crystallite Size (nm)	d-spacing (nm)
Ni NWs	44.33	0.42	22.0	2.04
Ni NTs	44.52	0.56	16.4	2.03
Co NWs	44.35	0.63	14.7	2.04
$Ni_{60}Fe_{40}$ NWs	44.41	0.48	19.3	2.04

Table 3. XRD peak positions, FWHM, crystallite size and d-spacing of various nanostructures.

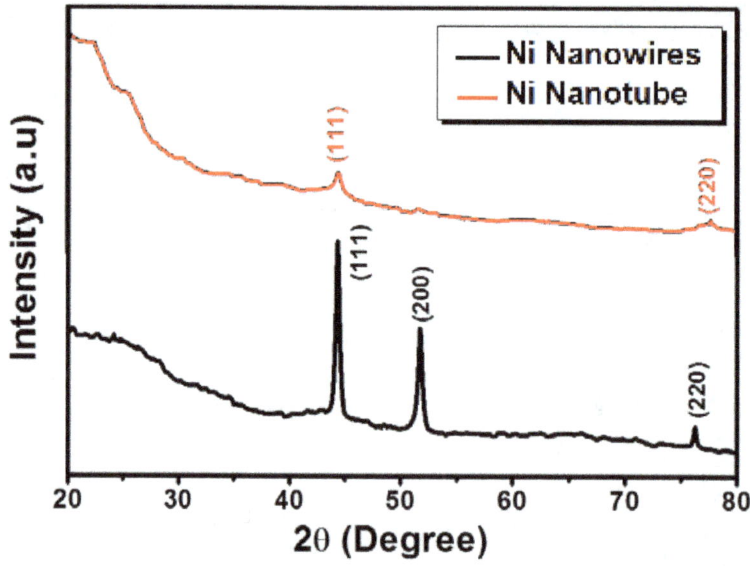

Figure 15. XRD spectra of Ni nanowires and nanotubes electrodeposited in AAO templates.

The hysteresis loops were measured in two geometries: when the applied magnetic field H is perpendicular to wire's axis and also parallel to wire's axis. The orientation of NWs and NTs embedded in AAO with applied external magnetic field in SQUID was ensured during sample preparation. For parallel and perpendicular arrangement, the templates were fixed in a small piece of folded straw which is further inserted in a new straw in such a way that it is along the easy and hard axis of straw elongation. An analysis of hysteresis loops allowed us to determine the values of coercivity (H_c) and normalized remanent magnetization (M_r). The saturation magnetization M_s value has been taken from reported works [42]. Figure 16 shows the hysteresis loop of typical Ni NWs having length 30 μm with the applied magnetic field parallel and perpendicular to the long axes of the nanowires. The difference between the hysteresis loops for the two orientations suggests the existence of magnetic anisotropy in the sample. The Ni NWs are not exactly uniform from the bottom to the top of the nanowires due to the Y-shaped templates being used. Y-shaped portion (Fig. 2) has a different shape anisotropy and dominates in magnetostatic coupling than the long-portion. This can only be seen as an averaged, smoothed curve in the SQUID measurements. Due to averaging, the easy axis loops show a curved shape and a consequent reduction in the squareness. The values of remanent magnetization $M_r(||)$ and $M_r(\perp)$ for 30 μm sample are 0.26 memu and 0.072 memu, respectively. This indicates that the magnetic easy axis of the system is along the axis parallel to the nanowires. The squareness (M_r/M_s) of the hysteresis curve is greater when the applied field is parallel to the nanowires than perpendicular to it. The values of coercivity $H_c(||)$ and $H_c()$ for 30 μm sample are 124 Oe and 84 Oe, respectively.

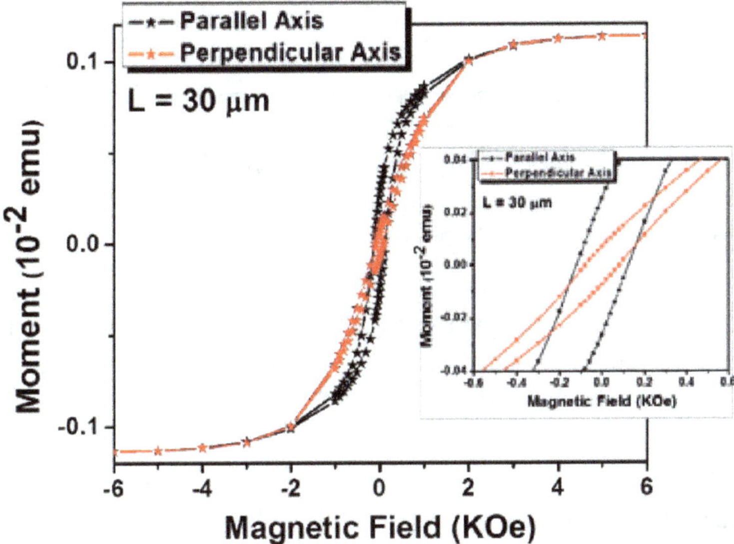

Figure 16. Magnetization loops for Ni NWs of length 30 μm at different orientation of the applied magnetic field at room temperature.

Figure 17 shows the magnetic hysteresis loop of Ni nanotube. The nanocylinders geometry, i.e. their length, inner and outer radii (R_{in} and R_{out}) and wall thickness (t_w) strongly affects the magnetization reversal mechanism. In our work also, we observed a magnetization reversal switching in nanotube geometry. One can easily observe that for nanowires (Fig. 16), the easy axis is along the wire's long axis, whereas for nanotubes (Fig. 17), the easy axis is parallel to the tube axis resulting in a magnetization switching. Since the deposited nanostructures have an *fcc* structure, which shows small magneto-crystalline anisotropy, magnetic anisotropy is mainly decided by the competition between the shape anisotropy of the individual NW/NT and the magnetostatic interaction between neighbouring NWs/NTs.

The values of coercivity $H_c(||)$ and $H_c(\perp)$ for NTs are 32 Oe and 41 Oe, respectively. Both curves are highly sheared, indicating strong inter-tubular interaction. Further, it is clear from the evidence that the array has very low remanence magnetization. The values of remanent

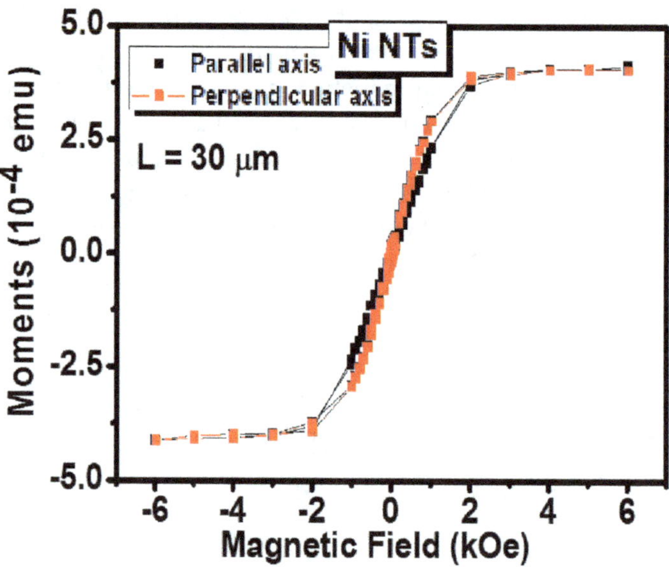

Figure 17. Magnetization loops for Ni NTs of length 30 μm at different orientation of the applied magnetic field at room temperature.

magnetization $M_r(\parallel)$ and $M_r(\perp)$ for 30 μm Ni NT sample are 0.006 memu and 0.017 memu, respectively. The remanent magnetization of Ni NTs is very small as compared to NWs, which is due to less volume of magnetic material present in nanotubes (hollow inside).

5. High-frequency applications

Investigations on the magnetization reversal modes of Ni NWs/NTs are done by angular-dependent FMR measurements. Here we will discuss the quantitative analysis of resonance frequency (f) and frequency line-widths (Δf) data. Depending upon the geometry of the nanostructures, basically three main modes of magnetization reversal exist: *coherent mode* (C), where all spins rotate homogenously; *vortex mode* (V), in which vortex-like domain wall nucleates and propagates; *transverse mode* (T), in which spins rotate progressively via propagation of a transverse domain wall.

Figure 18(A-D) shows the transmission response for arrays of Ni NWs and NTs used in the present study. In order to view the effect of interactions between NWs/NTs of FMR spectra, the resonance frequency and frequency line-width as a function of applied field was observed. Resonance frequency increases linearly with the increase of external magnetic field for all four samples. The magnetic bias field was rotated from parallel orientation of nanostructures to that of perpendicular to them for angular variation measurements. The resonance frequency ($f_r = \omega/2\pi$) as a function of field orientation can be obtained from Landau–Lifshitz–Gilbert equation as [21, 26]:

$$\left(\frac{\omega}{\gamma}\right)^2 = \left[Hcos\left(\theta - \theta_H\right) + H_{eff}cos^2\theta\right]\left[Hcos\left(\theta - \theta_H\right) + H_{eff}cos2\theta\right] \qquad (6)$$

where $\omega = 2\pi f$ and $\gamma = g\mu_B/\hbar$ is the gyromagnetic ratio, θ_0 is the angle between magnetization equilibrium direction and easy axis, and θ_H is the applied field direction measured from the easy axis. At the resonance frequency $\omega = \omega_r$, $\theta = \theta_H$ and the applied field $H = H_{res}$ in the saturated case.

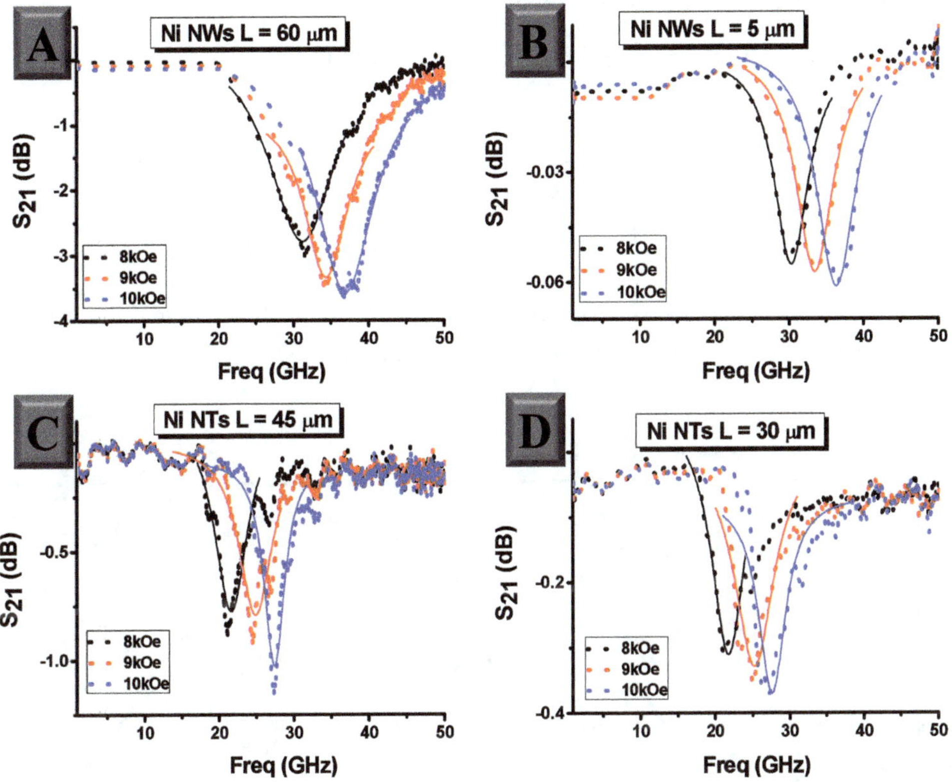

Figure 18. (A-D). Experimental transmission response (S_{21} vs. frequency) at different applied fields along parallel orientation of long axis of ferromagnetic nanowires and nanotubes of diameter 200 nm and various lengths. Solid lines are the Lorentzian fits to the experimental data.

The angular dependence of f_r for NWs and NTs is shown in Fig. 19(A). NW array exhibits stronger magnetic interaction than the NT arrays. In nanotubes, there is a transition in magnetization reversal mode at very small tube radius. The curling mode of magnetization reversal in an infinite cylinder predicts that f_r increases as angle (θ_H) increases from 0° to 90°, whereas coherent rotation mode gives highest and lowest f_r values for $\theta_H = 0°$ and $\theta_H = 90°$, respectively.

Figure 19(B) shows the theoretical results for angular dependence of f_r for NW and NT arrays. The results were calculated using Eq. (6) for the uniform mode and the corresponding equilibrium conditions. In a nanowire array, each nanowire is exposed to the field created by the neighbouring wires in addition to the self-demagnetizing field and hence the demagnetization field is given as $H_d = 2\pi M_s$ (1-3P), where P is the volume fraction of the magnetic nanostructure in the matrix. In case of nanotubes, the demagnetization field is no longer similar to that of the nanowires. Also, it is assumed that the nanotubes do not have the extreme surfaces as seen in wires to give rise to an additional interaction field that would oppose the tubes' shape anisotropy. It is also found that the magnetic anisotropy is sensitive to the demagnetization factor, demonstrating that the magnetostatic interaction between nanotubes is responsible for the wall-thickness-dependent magnetic anisotropy [43].

With this consideration, the demagnetization factors used to calculate H_{eff} for nanotubes are given as [44]:

$$N = \frac{1}{2}\left[1 - \lambda^2\left(\frac{\mu_r - 1}{\mu_r + 1}\right)\right] \tag{7}$$

where μ_r is the relative permeability and $\lambda = R_{in}/R_{out}$, where R_{in} is the inner and R_{out} is the outer radii of the nanotube, respectively. In our case λ is found to be 0.6.

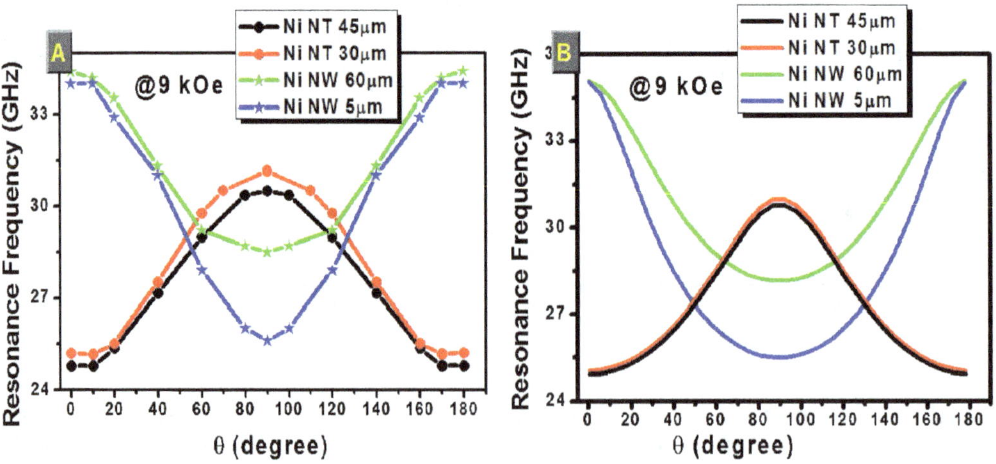

Figure 19. (A) Angular variation of the FMR frequency positions ($f_r(\theta_H)$) measured at an applied field of 9 kOe, (B) theoretical curves fitted from Eq. (6). $\theta_H = 0°$ corresponds to the applied magnetic field parallel to the NWs/NTs axis and $\theta_H = 90°$ corresponds to perpendicular configuration.

The bell-shaped f_r curves in Fig. 19(A) clearly shows a negative effective field for the nanotubes with the easy plane perpendicular to the nanotubes. At the parallel (low and high extreme values of θ_H) and perpendicular conditions there is a good agreement between the theoretical and experimental f_r value for both nanotubes and nanowires, while in the range of $\theta_H = 40°$-$80°$ and corresponding angles in the next quadrant, there is a difference in the f_r values. This difference can be attributed to the situation where the spins in the nanostructures under consideration vary spatially and hence the effective field cannot explain the mean magnetization in a particular direction giving rise to the coherent uniform rotation. This reveals the inhomogeneous internal and stray field in the NWs and NTs of small radii, which needs further investigation.

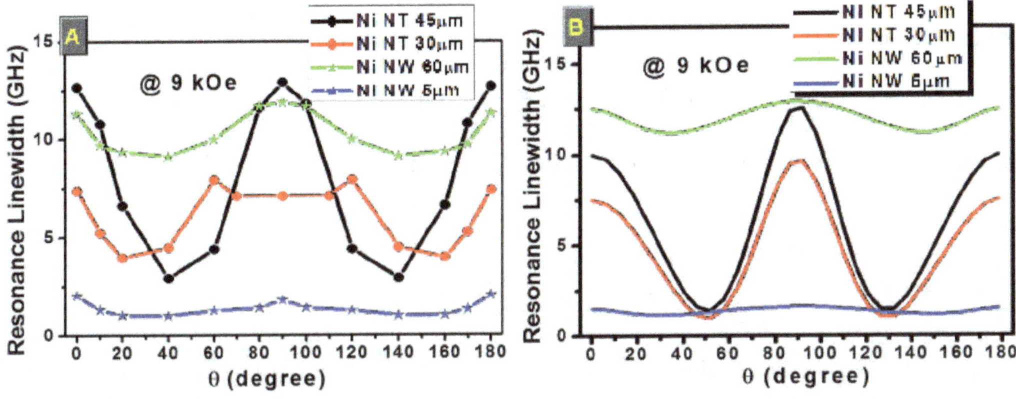

Figure 20. (A) Angular variation of FMR line-widths (Δf) measured for all samples at an applied magnetic field of 9 kOe, (B) theoretical curves fitted from Eq. (8).

Figure 20(A) shows the FMR line-width (Δf) as a function of θ_H for all samples. Δf was determined from the Lorentzian fit to the transmitted signal from the sample. The line-width reflects the distribution of the parameters of individual nanostructures that vary in their exact orientation inside the template as well as in their value of effective anisotropy field. Theoretical curves showing the angular dependence of frequency line-width is shown in Fig. 20(B). The frequency line-width is found using the following relation:

$$\Delta\left(\frac{\omega}{\gamma}\right) = \left(\frac{d(\omega/\gamma)}{dH}\right)\Delta H(\theta_H) \tag{8}$$

where ΔH is given by [45]

$$\Delta H = \Delta H_0 + 1.16\alpha\left(\frac{\omega}{\gamma}\right) \tag{9}$$

ΔH_0 is the frequency independent contribution to the line-width caused by inhomogeneous broadening. It is found that the shape of the line-width curves is similar in both experiment and theory.

These studies show that ferromagnetic nanostructures embedded in AAO templates fabricated by electrodeposition technique are the best candidates for high-frequency application devices.

Author details

Monika Sharma[1*] and Bijoy K. Kuanr[2]

*Address all correspondence to: monikasharma1604@gmail.com

1 Indian Institute of Technology Delhi, New Delhi, India

2 Special Centre for Nanoscience, Jawaharlal Nehru University, New Delhi, India

References

[1] F. Byrne, A. Prino-Mello, A. Whelan, B. M. Mohamed, A. Davis, Y. K. Gunko, J. M. D. Coey, and Y. Volkov, "High content analysis of the biocompatibility of nickel nanowires, " *J Magn Magn Mater*. 321 (2009), pp. 1341-1345.

[2] A. A. Stashkevich, Y. Roussigne, P. Djemia, S. M. Cherif, P. R. Evans, A. P. Murphy, W. R. Hendren, R. Atkinson, R. J. Pollard, A. V. Zayats, G. Chaboussant, and F. Ott, "Spin-wave modes in Ni nanorod arrays studied by Brillouin light scattering, " *Phys Rev B*. 80 (2009), pp. 144406.

[3] D. V. Berkov, C. T. Boone and I. N. Krivorotov, "Micromagnetic simulations of magnetization dynamics in a nanowire induced by a spin-polarized current injected via a point contact, " *Phys Rev B*. 83 (2011), pp. 054420.

[4] S. F. Chen, H. H. Wei, C. P. Liu, C. Y. Hsu, and J. C. A. Huang, "Microstructural effects on the magnetic and magnetic-transport properties of electrodeposited Ni nanowire arrays, " *Nanotechnology*. 21 (2010), pp. 425602.

[5] A. N. Banerjee, S. Qian, and S. W. Joo, "Large field enhancement at electrochemically grown quasi-1D Ni nanostructures with low-threshold cold-field electron emission, " *Nanotechnology*. 22 (2011), pp. 035702.

[6] X. Kou, X. Fan, R. K. Dumas, Q. Lu, Y. Zhang, H. Zhu, X. Zhang, K. Liu and J. Q. Xiao, "Memory effects in magnetic nanowire arrays", *Adv Mater*. 23 (2011), pp. 1393-1397.

[7] Y. Xia, P. Yang, Y. Sun, Y. Wu, Y. Yin, F. Kim, and H. Yan, "One-dimensional nanostructures: Synthesis, characterization and applications, " *Adv Mater*. 15 (2003), pp. 353-389.

[8] Y. C. Sui, R. Skomski, K. D. Sorge, and D. J. Sellmyera, *Appl Phys Lett.* 84 (2004), pp. 1525.

[9] X. Duan, and C.M. Lieber, "General synthesis of compound semiconductor nanowires, " *Adv Mater.* 12(2000), pp. 298–302.

[10] T. M. Whitney, J. S. Jiang, P. C. Searson, and C. L. Chien, "Fabrication and magnetic properties of arrays of metallic nanowires, " *Science.* 261 (1993), pp. 1316-1319.

[11] N. J. Gerein, and J. A. Haber, "Effect of ac electrodeposition conditions on the growth of high aspect ratio copper nanowires in porous aluminium oxide templates, " *J Phys Chem B.* 109 (2005), pp. 17372-17385.

[12] S. Yue, Y. Zhang, and J. Du, "Preparation of anodic aluminium oxide tubular membranes with various geometries, " *Mater Chem Phys.* 128 (2011), pp. 187-190.

[13] F. Tao, M. Guan, Y. Jiang, J. Zhu, Z. Xu, and Z. Xue, "An easy way to construct an ordered array of nickel nanotubes: the triblock-copolymer-assisted hard-template method, " *Adv Mater.* 18 (2008), pp. 2161-2164.

[14] X. W. Wang, Z. H. Yuan, S. Q. Sun, Y. Q. Duan and L. J. Bie, "Electrochemically synthesis and magnetic properties of Ni nanotube arrays with small diameter, " *Mater Chem* Phys. 112 (2008), pp. 329-332.

[15] D. H. Park, Y. B. Lee, M. Y. Cho, B. H. Kim, S. H. Lee, Y. K. Hong, J. Joo, H. C. Cheong, S. R. Lee, "Fabrication and magnetic characterization of hybrid double walled nanotube of ferromagnetic nickel encapsulated conducting polypyrrole, " *Appl Phys Lett.* 90 (2007), pp. 093122.

[16] J. Bao, C. Tie, Z. Xu, Q. Zhou, D. Shen, and Q. Ma, "Template synthesis of an array of nickel nanotubules and its magnetic behaviour, " *Adv Mater.* 13 (2001), pp. 1631-1633.

[17] J. Podbielski, F. Giesen, M. Berginski, N. Hoyer, and D. Grundler, "Spin configurations in nanostructured magnetic rings: From DC transport to GHz spectroscopy, " *Superlattices & Microstructures.* 37 (2005), pp. 341-348.

[18] S. Sun, C. B. Murray, D. Weller, L. Folks, and A. Moser, *Science.* 287 (2000), pp. 1989.

[19] J. Escrig, M. Daub, P. Landeros, K. Nielsch, and D. Altbir, "Angular dependence of coercivity in magnetic nanotubes, " *Nanotechnology.* 18 (2007), pp. 445706.

[20] A. Encinas, M. Demand, L. Vila, L. Piraux, and I. Huynen, "Tunableremanent state resonance frequency in arrays of magnetic nanowires, " *Appl Phys Lett.* 81 (2002), pp. 2032.

[21] M. Sharma, B. K. Kuanr, M. Sharma and A.Basu, "Relation between static and dynamic magnetization effects and resonance behaviour in Ni nanowire arrays, " *IEEE Trans Magn.* 50 (2014), pp. 4300110.

[22] X. P. Huang, W. Han, Z. L. Shi, D. Wu, M. Wang, R. W. Peng, andN. B. Ming, "Electrodeposition of periodically nanostructured straight cobalt filament arrays, " *J Phys Chem.* C 113 (2009), pp. 1694–1697.

[23] S. Dubois, J. M. Beuken, L. Piraux, J. L. Duvail, A. Fert, J. M. George, and J. L. Maurice, *J Magn Magn Mater.* 165 (1997), pp. 30.

[24] M. Sharma, B. K. Kuanr, M. Sharma and A. Basu, "New opportunities in microwave electronics with ferromagnetic nanowires, " *J Appl Phys.* 115 (2014), pp. 17A518.

[25] L. P. Carignan, T. Kodera, A. Yelon, C. Caloz, and D. Menard, "Integrated and self-biased planar magnetic microwave circuits based on ferromagnetic nanowire substrates, " *Microwave Conf.* (2009), pp. 743-746.

[26] R. L. Marson, B. K. Kuanr, S. R. Mishra, R. E. Camley and Z. Celinski, "Nickel nanowires for planar microwave circuit applications and characterization, " *J Vac Sci Tech B.* 25 (2007), pp. 2619.

[27] A. Huczko, "Template-based synthesis of nanomaterials, " *Appl Phys A.* 70 (2000), pp. 365.

[28] N. I. Kovtyukhova, B. R. Martin, J. K. N. Mbindyo, T. E. Mallouk, M. Cabassi, and T. S. Mayer, *Mater Sci Engin C.* 19 (2002), pp. 255.

[29] R. C. Furneaux, W. R. Rigby, and A. P. Davidson, "The formation of controlled-porosity membranes from anodically oxidized aluminium, " *Nature.* 337 (1989), pp. 147-149.

[30] R. L. Fleisher, P. B. Price, and R. M Walker, *Nuclear Tracks in Solids,* Univ. of California Press, Berkeley, (1975), pp. 562-595.

[31] A. Despic, and V. P. Parkhuitik, *Modern Aspects of Electrochemistry,* vol. 20 (1989), Plenum, New York, pp. 401.

[32] T. Nevin, O. Sadullah, K. Necmettin, Y. Hayrettin, and O. Z. Zafer, "Simple fabrication of hexagonally well-ordered AAO template on silicon substrate in two dimensions, " *Appl Phys A.* 95 (2009), pp. 781-787.

[33] H. Chik, and J. M. Xu, "Nanometric superlattices: non-lithographic fabrication, materials, and prospects, " *Mater Sci Engin R.* 43 (2004), pp. 103-138.

[34] A. P. Li, F. Muller, and U. Gosele, "Polycrystalline and Monocrystalline Pore Arrays with Large Interpore Distance in Anodic Alumina, " *Electrochem. Solid State Lett.* 3 (2000), pp. 131.

[35] A. J. Bard, and L. R. Faulkner, *Electrochemical Methods: Fundamentals and Applications,* Wiley: New York (1980).

[36] M. P. Proenca, C. T. Sousa, J. Ventura, M. Vazquez, and J. P. Araujo, "Ni growth inside ordered arrays of alumina nanopores: Enhancing the deposition rate, " *Electrochim Acta.* 72 (2012), pp. 215-221.

[37] M. P. Proenca, C. T. Sousa, J. Ventura, M. Vazquez, and J. P. Araujo, "Distinguishing nanowire and nanotube formation by the deposition current transients, " *Nano Res Lett.* 7 (2012), pp. 280.

[38] T. N. Narayanan, M. M. Shaijumon, Lijie Ci, P. M. Ajayan, and M. R. Anantharaman, "On the growth mechanism of nickel and cobalt nanowires and comparison of their magnetic properties, " *Nano Res.* 1 (2008), pp. 465-473.

[39] G. Ali, M. Ahmad, J. I. Akhter, M. Khan, K. Shafqat, M. Maqbool, and S. G. Yang, "Characterization of cobalt nanowires fabricated in anodic alumina template through AC electrodeposition, " *IEEE Trans Nanotech.* 9 (2010), pp. 223.

[40] K. Y. Kok, C. Hangarter, B. Goldsmith, I. K. Ng, N. U. Saidin, and N. V. Myung, "Template assisted growth and characterization of electrodeposited permalloy (Ni80Fe20)/Cu multi-layered nanowires, " *ECS Trans.* 25 (2010), pp. 97-103.

[41] H. P. Klug, and L. E. Alexander, *X-ray Diffraction Procedures*, Wiley, New York, pp. 125 (1962).

[42] D. J. Sellmyer, M. Zheng, and R. Skomski, "Magnetism of Fe, Co and Ni nanowires in self-assembled arrays, " *J Phys Cond Matt.* 13 (2001), pp. R433-R460.

[43] Y. L. Li, S. L. Tang, R. Xie, Y. Wang, M. Yang, J. L. Gao, W. B. Xia, and Y. W. Du, "Fabrication and magnetic properties of free-standing Ni nanotube arrays with controllable wall thickness, " *Appl Phys Lett.* 100 (2012), pp. 052402.

[44] J. Prat-Camps, C. Navau, D. X. Chen, and A. Sanchez, "Exact analytical demagnetizing factors for long hollow cylinders in transverse field, " *IEEE Magn Lett.* 3 (2012), pp. 0500104.

[45] B. Heinrich, J. F. Cochran, and F. R. Hasegawa, "FMR line broadening in metals due to two-magnon scattering, " *J Appl Phys.* 57 (1985), pp. 3690.

[46] B. Bhushan, *Springer Handbook of Nanotechnology*, Chapter -5 by H. (Mary) Shang and G. Cao, "Template-based synthesis of nanorod or nanowire arrays, " Springer (2010), pp. 169-186.

2

Electrodeposition of Nanostructure Materials

Souad A. M. Al-Bat'hi

Abstract

We are conducting a multi-disciplinary research work that involves development of nanostructured thin films of semiconductors for different applications. Nanotechnology is widely considered to constitute the basis of the next technological revolution, following on from the first Industrial Revolution, which began around 1750 with the introduction of the steam engine and steelmaking. Nanotechnology is defined as the design, characterization, production, and application of materials, devices and systems by controlling shape and size of the nanoscale. The nanoscale itself is at present considered to cover the range from 1 to 100 nm. All samples prepared in thin film forms and the characterization revealed their nanostructure. The major exploitation of thin films has been in microelectronics, there are numerous and growing applications in communications, optical electronics, coatings of all kinds, and in energy generation. A great many sophisticated analytical instruments and techniques, largely developed to characterize thin films, have already become indispensable in virtually every scientific endeavor irrespective of discipline. Among all these techniques, electrodeposition is the most suitable technique for nanostructured thin films from aqueous solution served as samples under investigation. The electrodeposition of metallic layers from aqueous solution is based on the discharge of metal ions present in the electrolyte at a cathodic surface (the substrate or component.) The metal ions accept an electron from the electrically conducting material at the solid- electrolyte interface and then deposit as metal atoms onto the surface. The electrons necessary for this to occur are either supplied from an externally applied potential source or are surrendered by a reducing agent present in solution (electroless reduction). The metal ions themselves derive either from metal salts added to solution, or by the anodic dissolution of the so-called sacrificial anodes, made of the same metal that is to be deposited at the cathode.

Keywords: Electrodeposition, nanomaterials, thin films, nanotechnology

1. Introduction

Materials with grain size less than 100 nm are classified as nanostructure materials [1, 2]. Due to the ultra-small building units and high surface/volume ratio, these materials exhibit special properties, such as mechanical, optical, electronic, and magnetic properties [3]. In nanostructure materials, all size-related effects can be integrated by controlling the sizes of the constituent components [4]. For instance, nanostructure metals and ceramics can have improved mechanical properties compared to conventional metals and ceramics. Besides, nanostructure materials are capable of being sintered at much lower temperatures than conventional powders, enabling the full densification of these materials at relatively lower temperatures. Technological applications of semiconductor nanostructure materials in optoelectronic devices such as photodiodes and quantum dotes semiconductors are owed to the size-related effects, particularly quantum size effects caused by the spatial confinement of delocalized electrons in confined grain sizes [5]. Nanostructure materials magnetic applications are fabrication of devices with massive magnetoresistance (GMR) effects, to read data on computer hard drives by magnetic heads, and development of magnetic refrigerators by replacing compressed ozone-chlorofluorocarbons by solid magnets as refrigerants [6]. Nanostructured metals seem to be a candidate for new catalytic applications [7].

Fundamental advances in energy conversion and storage which are full of vigor in meeting outfaces of some environmental phenomena such as global warming and impact of fossil fuels are held by new materials, particularly nanomaterials, which present unique properties as electrodes and electrolytes in many applications such as lithium batteries, solar cells, fuel cells, and supercapacitors [8].

There are several techniques that can be used to develop thin films of metals and semiconductors in both micro- and nano-structures. Some of these techniques are: sol-gel process [9], chemical deposition method [10], hydrothermal method [11], pyrolysis method [12], chemical vapor deposition (CVD) [13], and electrodeposition method [14]. In Table 1, we summarize some previous studies on general concepts of nanostructure materials, their properties, and their methods of preparation. This chapter will be concerned particularly with the electrodeposition technique. Figure 1 shows thin films deposition techniques.

Materials may undergo some changes in physical properties when changing state from microstructure to a nanostructure. High surface/volume ratio and particles' transfer to a domain with dominant quantum effects are two main factors in characteristics' change. An increase in surface/volume ratio results in the dominance of surface atoms' behavior compared with internal atoms. This factor affects both the particle's properties and its interaction with the other materials. Vast surface of the nanoparticle leads to specific features such as mechanical, chemical, and thermal properties.

Nanoparticles are the basic structural blocks in nanotechnology. They are the starting point in producing nanostructure materials and tools. The production of nanoparticle is an important factor in the development of research in nanoengineering and nanoscience.

Different types of nanoparticles are prepared by using precursors in solid, liquid, and gaseous phases in the production and arrangement of nanomaterials. Based on their chemical reactivity or physical compression, these materials align the nanostructures as structural blocks side-by-side.

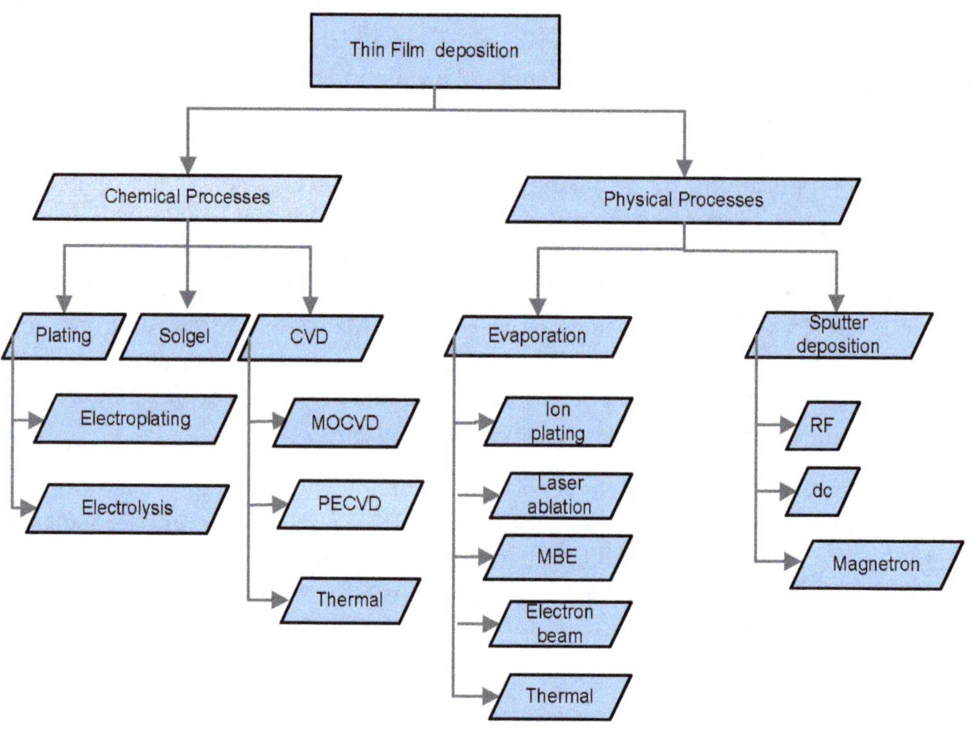

Figure 1. Thin film deposition techniques [Souad A., (2007)] [15].

	Title	Description	Reference
1.	Electrodeposition of Copper Oxide Nanoparticles on Precasted Carbon Nanoparticles Film for Electrochemical Investigation of anti-HIV Drug Nevirapine	A novel medical application is developed for the analysis of the anti-HIV drug, nevirapine (NEV), based on electrodeposition of CuO/CNP nanoparticles. The modified electrode (sensor) has shown a good response, which makes it suitable for the determination of NEV in	Saeed, S. et al., Electroanalysis. DOI: 10.1002/elan. 201500027,2015, WILEY-VCH [16]

	Title	Description	Reference
		real samples of human blood serum.	
2.	Electrochemical Synthesis of Mesoporous CoPt Nanowires for Methanol Oxidation	Thin films CoPt nanowires have been successfully synthesized by electrodeposition technique. Electrodeposition helps easy synthesis of mesoporous materials useful for catalytic applications without any agent's employment. The mesoporous CoPt nanowires have shown enhanced catalytic behavior for methanol oxidation reaction in acidic medium. Because of the catalytic activity and cost reduction, these nanowires can be a potential alternative for methanol fuel cells.	Albert S. et al., Nanomaterials (2014), 4(2), 189-202 [17]
3.	Nanostructured Solids From Nano-Glasses to Quantum Transistors Materials Assembled from Atom Clusters	A review provides a discussion of method of preparation, structure and properties of nano structured solids that consist of totally crystalline, totally amorphous, and partially amorphous components.	H. Gleiter, Th. Schimmel, H. Hahn, Nano Today.. (2014) 9, 17-68 [18]
4.	Nanocarbon-Based Electrochemical Systems for Sensing, Electrocatalysis, and Energy Storage	A review provides comparison between carbon nanofibers, carbon nanotubes, and graphene; the three types of carbon nanostructure materials which have been focused on.	Xianwen et al., Nano Today. (2014) 9, 405-432 [19]
5.	Electrodeposition of Polypyrrole/ Reduced Graphene Oxide/Iron Oxide Nanocomposite as Supercapacitor Electrode Material	Nanocomposite polymer thin films prepared by electrodeposition technique have shown an enhanced electrochemical stability and exhibit excellent specific capacitance (F/gm)	Eeu Y. C. et al., Hindawi Publishing Corporation J Nanomater. Volume 2013, (2013), Article ID 653890 [20]

	Title	Description	Reference
6.	ZnxCd1-x (O) Thin Film Nanorods for PV Applications	A hexagonal nanorods thin film of ZnxCd1-x (O) were electrodeposited on ITO substrate and is used as working electrode in photovoltaic cells to generate electricity from solar energy by absorbing sun light.	Umer M. and, Souad. A., Int J Nano Sci Technol. (2013) 1, (3), April, 09-16 [21]
7.	Electrodeposition of Platinum and Ruthenium Nanoparticles in Multiwalled Carbon Nanotube-Nafion Nanocomposite for Methanol Electrooxidation	A 10-15 nm diameter of PtRu nanoparticle have been electrodeposited within nanocomposite (MWCNT-Nafion). cyclic voltammetry is used to investigate the electrocatalytic activity toward the methanol electrooxidation at three different nanocomposite electrodes.	Yu-Huei H. and Yu Chen T., J Nanomater. Volume 2009 (2009) [22]
8.	Electrodeposition of One-Dimensional Nanostructures	Electrodeposition technique is successfully used to synthesize 1D nanostructure materials with and without temples and a comparison between the two approaches is well discussed.	Guangwei S., Lixuan M., and Wensheng S., Recent Patents Nanotechnol. (2009), 3,182-191 [23]
9.	Characterization of Electrodeposited Zn (Se, Te) Thin Films/Polymer (PEO–Chitosan Blend) Junction for Solar Cells Applications	The work has focused on the possibility of using Zn (Te, Se) semiconductors with PEO-chitosan blend polymer electrolytes for PV applications. The methods to produce the semiconductors (electrodeposition) and polymer electrolyte (solution casting) are suitable and cheap. The economy of the processes involved such as the electrodeposition method and the solution cast technique, show some	Souad A. (2007), PhD thesis [15]

	Title	Description	Reference
		promise for future large-scale production.	
10.	Ionic Liquids as Green Electrolytes for the Electrodeposition of Nanomaterials	Air and water as ionic liquids attract the attention of researchers due to their amazing physical properties, particularly in electrodeposition of nanocrystalline metals. Copper and aluminum can be electrodeposited in the air and water ionic liquids on solid electrodes and they have shown an adequate electronic conductivity.	S. Zein El Abedin, et al., Green Chem. (2007),, 9, 549-553 [24]

Table 1. Summary of Some Previous Studies on General Concepts of Nanostructure Materials

2. Growth technique of nanomaterials thin films

Nanomaterials are produced by two main technical methods: "bottom-up" and "top-down" techniques. In the bottom-up technique, the substance developed from the bottom (atom by atom, molecule by molecule, or cluster by cluster). Colloid dispersion is produced in this technique. The top-down technique begins with bulk material and progresses toward the ideal state by its designing or abrasion. This technique is similar to the electron beam lithography in which patterns are used. In nanotechnology industry, bottom-up and top-down techniques are playing a considerable role. The major advantage of both techniques is creation of high-purity ultra-small structures. However, using bottom-up method produces improved nano-structures such as less or defect-free, homogeneous and long- and short-term orders, all due to free Gibbs energy decreases resulting in thermal balanced nanomaterials. However, in the top-down technique, mostly the material suffers from an increase in surface defects, since the material is subjected to internal stress [25]. Figure 2 shows a schematic representation of the building up of nanostructures.

3. Electrodeposition technique

Electrochemical deposition, or electrodeposition for short, refers to a film growth process which consists in the formation of a metallic coating onto a base material (substrate) occurring through the electrochemical reduction of metal ions from an electrolyte to achieve the desired electrical and corrosion resistance, reduce wear and friction, improve heat tolerance, and for

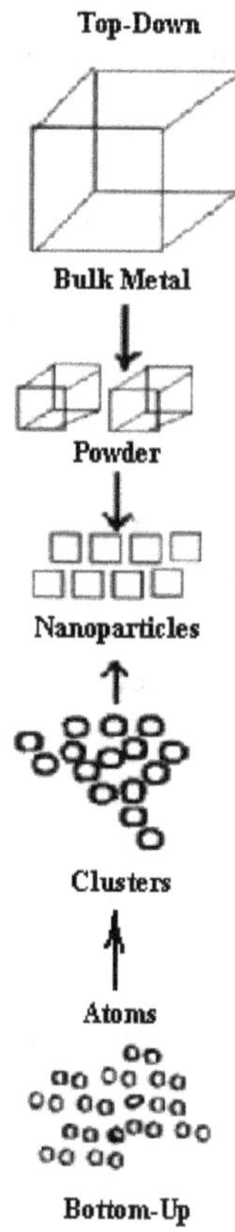

Figure 2. Schematic representation of the building up of nanostructures

decoration. The analogical technology is oftentimes known as electroplating. In addition to the production of metallic coatings, it is also for electrometallurgy in which metals are extracted from their ores or for electroformation where molds to form objects in their final shapes are produced. Mostly, the metallic deposit obtained is crystalline. This process can therefore be called electrocrystallization, a term introduced by russian chemist V. Kistiakovski in the early twentieth century.

The electrolyte contains positive and negative ions. Therefore, it is considered an ionic conductor. To prepare an electrolyte, the desired metal contained in a chemical species liquidized (mostly dissolved in water) to form a molten salt besides different organic and ionic liquids are currently used for particular electroplating processes. To begin electrodeposition, the cathode (working electrode, W.E) immersed in the electrolyte contained in a vessel (cell) along with the anode (counter electrode, C.E). To allow electric current flow in the circuit, the two electrodes are connected to a battery or any other power source. The cathode is connected to the negative terminal of the battery, while the anode is connected to the positive terminal so that the metal ions are reduced to metal atoms, which eventually form the deposit on the surface. Figure 3 shows a schematic presentation of an electrolytic cell for electroplating or electrodepositing a metal "M" from an aqueous (water) solution of metal salt "MA".

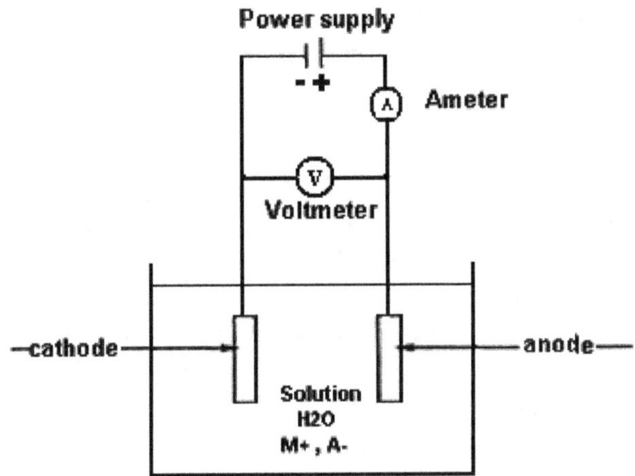

Figure 3. A schematic electrodeposition technique

This type of circuit arrangement directs electrons into a bath from the power supply to the cathode. In the bath, the electric current is carried by the positively charged ions from the anode toward the negatively charged cathode. This enables the metal ions in the bath to migrate toward extra electrons that are located at or near the surface of the cathode. Finally, the metal ions are removed from the solution and are deposited on the surface of the object as a thin layer. The reaction in aqueous medium at the cathode therefore obeys equation 1:

$$M^{+n} + ne^- \rightarrow M \tag{1}$$

where n is the number of electrons involved in the reaction.

The thickness of the electrodeposited layer on the substrate is determined by the time duration of the plating. The longer the object remains in the chemical bath, the thicker the deposited layer is [15].

Lately, electrodeposition has become increasingly popular and is being extensively used by researchers in disparate fields of inquiry. What exactly is causing such an interest? The process itself is deceptively simple, a conductive surface is immersed into an electrolyte containing ions of the material to be deposited and a voltage is applied across this solid/electrolyte interface, resulting in a charge transfer reaction and film deposition. The driving force for this process is the applied potential, a quantity that can be easily and precisely controlled down to the mV and over timescales as short as 1ns. This feature leads to considerable control over the material formation process, its microstructure, and properties. The challenge in developing electrodeposition processes today is not the synthesis of a predefined material, but to strike a compromise between the ideal conditions used to produce this material and the commercial feasibility of the process. Other often-cited advantages of electrodeposition include the high utilization rate of the raw materials, low energy consumption, low material waste, little capital investment, and ease of implementation. Although the latter two are usually considered an advantage, they are perhaps a double-edged sword, they may in fact encourage quick experimentation, thereby relegating the science to the backstage and undermining the efforts of the community to place this technology on par with vacuum/gas/plasma deposition methods.

New materials and processes are being developed and integrated in microelectronic manufacturing to provide new functionalities, for example, in MEMS, lab-on-a-chip, or microfluidic devices. Polymers and biomaterials are electrodeposited for biomedical applications; metal oxides and compound semiconductors are grown electrochemically for electronic or optoelectronic applications. Electrodeposition in new electrolytic media such as ionic liquids or supercritical fluids is being strongly pursued with some success. Electrodeposition plays an important role in the development of sustainable energy conversion technologies, both at the portable and on a global scale. In addition, the principles governing the scale-up of electrodeposition processes are well understood, facilitating the development of large-scale manufacturing processes [15].

3.1. The electrolysis cell

3.1.1. Constituents of the cell

The electrolysis process consists of four parts:

1. The external circuit, with power supply or battery as a source of direct current (dc), that is conveyed to the plating tank, and associated with instruments such as ammeters, voltmeters, and voltage and current regulator to regulate them at appropriate values.

2. The working electrodes or cathodes (the material to be plated). The cathodes are to be immersed in the plating solution so that contact is made with the current source.

3. The plating solution or solution bath.

4. The counter electrodes or anodes (metal or conducting materials being plated which serves merely to complete the circuit).

The electrodeposition cell containing the plating solution or the electrolyte is usually made of plain mild steel (for alkaline solutions) or of lined steel (for acidic solutions). Steel lining could be of rubber, plastics, glass, or even lead [26].

3.1.2. Electrolytes (Solution baths)

The solution bath serves as a source of metal to be deposited, which contains ions of that metal, hence it provides conductivity, a stabilizer to stabilize the solution, for example, against hydrolysis and to stabilize the solution pH when acts as a buffer, regulates the corporal form of the deposit and helps dissolve the anodes. Finally, modifies properties of either the solution or of the deposit,

Complex ions are formed to stabilize the cation, which becomes much more stable when complexed to some ligand or to metal atoms by coordinate bonding. The concentration of the free (equated) ion is lowered by the presence of the ligand. Complex ions are also formed to hold the equated form at suitably low concentration allowing control of the evenness of plating.

Electrolytes can be defined as substances such as acid, base, and salts, which can conduct electricity in their aqueous solution due to ionization. The ions existing in the electrolyte are responsible for the conduction that results in current or free electrons passing through the wires. When a voltage is applied across the cell, and under the influence of the potential difference, the negatively charged ions migrate to the anode and the positively charged ions migrate to the cathode. The flowing of ions in aqueous solution of an electrolyte is called electrolysis [27]. For electrolysis of KCl in aqueous solution, it must be molten first by heating and elevating its temperature to the melting point. The molten ionic compound will be dissociated by electrolysis process into its elements. The reaction at each electrode is shown in equations 2 and 3:

$$2K^+ + 2e^- \rightarrow 2K_{(M)} \left(\text{At the Cathode} \right) \tag{2}$$

$$2Cl^- - 2e^- \rightarrow Cl_{2(g)} \left(\text{At the Anode} \right) \tag{3}$$

At the cathode, potassium ions are reduced to potassium atoms (metal) by gaining electrons. On the other hand, at the anode, chloride ions are oxidized to chlorine atoms by losing electrons which are combined to form molecules of chlorine gas.

The combining of equations 2 and 3 results in the following reaction:

$$2K^+Cl^-_{(l)} \rightarrow 2K_{(s)} + Cl_{2(g)} \left(\text{Overall Reaction} \right) \tag{4}$$

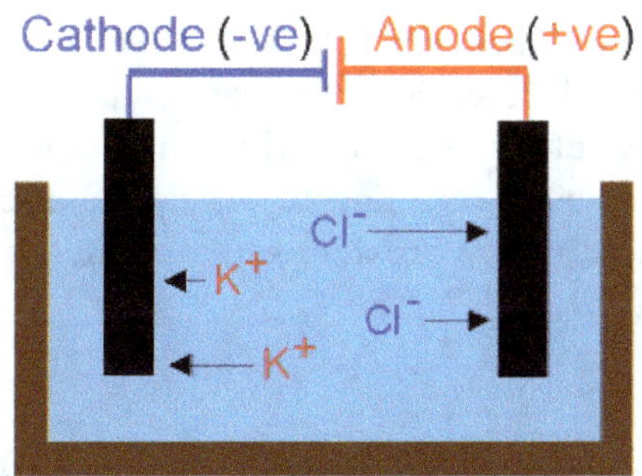

Figure 4. Electrolysis process of potassium chloride [28].

Strong Electrolytes

Some electrolytes such as potassium chloride, sodium hydroxide, and sodium nitrate are completely ionized in their constituent ions (>99%) in solution. These electrolytes are known as strong electrolytes. In other words, strong electrolytes are 100% dissociated in their aqueous solution. Generally, strong acids like sulfuric acid, nitric acid, hydrochloric acid; strong bases like potassium hydroxide salt and many others are strong electrolytes [27]. Salts that are composed of oppositely charged ions held by strong electrostatic forces of attractions are considered strong electrolytes and they dissolve in water, which highly weakens the attraction forces between ions and reduces it by a factor of 80 due to its high dielectric constant of 80. Therefore, it facilitates the free movement of ions that are stabilized by solvation/hydration process with water. Figure 5 demonstrates how sodium and chloride ions are surrounded by water molecules due to ion-dipole interaction.

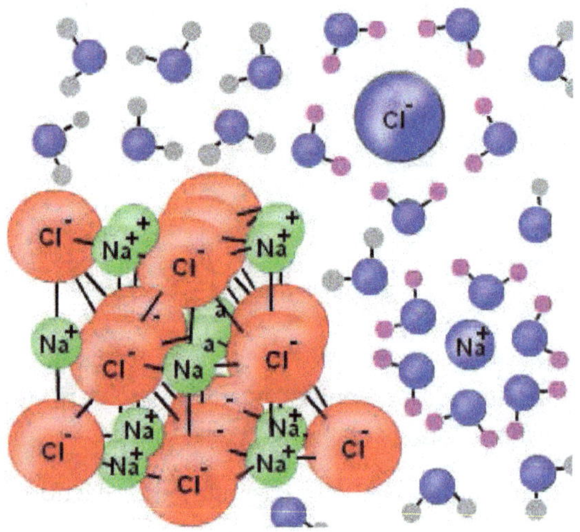

Figure 5. Sodium chloride solvation process

Weak Electrolytes

Some electrolytes are weak. A weak electrolyte is an electrolyte that partially dissociates in solution. In other words, ions and molecules of the electrolyte are both included in the solution. A moderately small percentage of ions (<1%) are given in solution with most of the compound staying in a non-ionic form (>99%) resulting in poorly conducting solutions. Acetic acid, carbonic acid, ammonia, rubidium hydroxide, some organic bases, and many others are considered weak electrolytes [29].

Nonelectrolytes

Substances that do not dissociate into ions in solution (water) are known as nonelectrolytes. These substances include most carbon compounds such as methyl alcohol, ethyl alcohol, glucose, and many others [30].

3.1.3. Electrode potential

When a metal electrode is placed in an ionic solution, ions will be exchanged between the metal and the solution. Ions from the metal enter the solution, and ions from the solution enter the metal lattice. The boundary between the two phases is called the interface. Due to an electron exchange between the electrode and the ions in solution, the interface becomes electrified and a potential difference across the metal-solution interface is generated. This potential difference is called electrode potential.

In case more ions leave than enter the metal electrode, there is an excess of electrons on the metal and hence it acquires negative charge, which attracts positively charged metal ions from the solution and repels negatively charged ions. Therefore, an excess of positive ions in the solution near the metal interface and the solution side acquires opposite (positive) and equal charge. The rate of positive ions leaving the metal electrode slows down due to repulsion with the positive charge at the solution side of the interface and the rate of ions entering the metal electrode accelerates [31]. In a short time, equilibrium between the metal and its ions in the solution is achieved, as shown in Figure 6.

$$\vec{n} = \tilde{n} \tag{5}$$

where, $\vec{n}$ denotes the number of positive ions that enter the metal electrode and $\tilde{n}$ denotes the number of positive ions that leave the metal electrode.

At equilibrium:

$$q_M = -q_S \tag{6}$$

where, q is the electric charge.

The result of the changing of the interface is the potential difference, $\Delta\Phi$ (M, S), between the potentials of the metal, Φ_M, and the solution, Φ_S:

$$\Delta\Phi(M, S) = \Phi_M - \Phi_S \tag{7}$$

In order to measure the potential difference of an interphase, one must connect it to another one and thus form an electrochemical cell. The potential difference across this electrochemical cell can be measured [32].

The cell shown in Figure 7 may be schematically represented in the following way:

$$Pt, \; H_2\left(p=1\right) \big| H^+\left(a=1\right) \big\| Cu^{2+}\left(a=1\right) \big| Cu \big| Pt \big|$$

where the electrode at the left side is the hydrogen reference electrode. When the pressure of H_2 p = 1 atm and the activity of H^+ ions a = 1, the hydrogen electrode is called the standard hydrogen electrode (SHE) and its potential is zero by convention. The standard electrode potential of a solution, E° (V) is the tendency for a chemical species to be reduced. The more positive the potential, is the more likely it will be reduced [1].

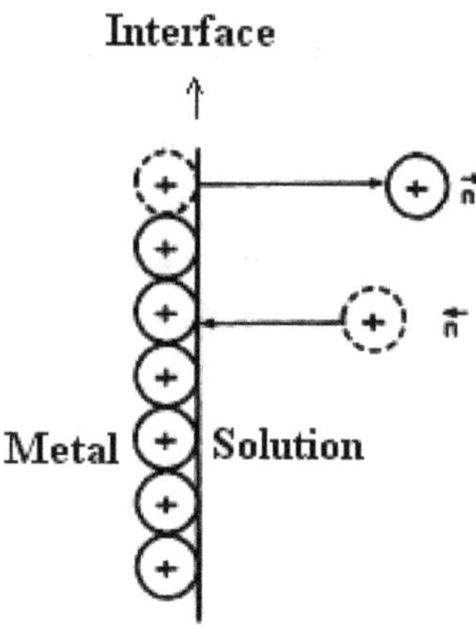

Figure 6. Formation of metal-solution interface (equilibrium state, $\vec{n}=\bar{n}$)

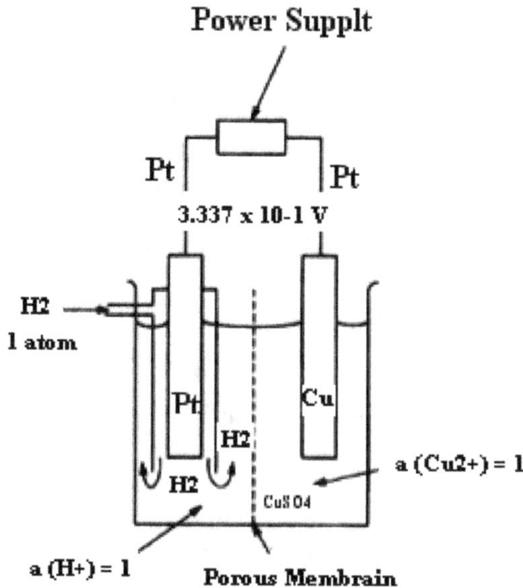

Figure 7. Relative standard electrode potential E^0 of a Cu/Cu^{2+} electrode

3.1.4. Some electrode reactions

Some acidic and basic solution reductions with their corresponding standard electrode potential E^0 measured at 25ºC are presented. Obviously, the more positive the potential E^0 is, the more likely it will be reduced. However, the potential E^0 of the reference electrode is zero, below which the potential E^0 is increasing negatively, as shown below [33]:

$F_2(g) / F^-(aq)$	+2.87
$Co^{3+}(aq) / Co^{2+}(aq)$	+1.82
$Pb^{4+}(aq)/Pb^{2+}(aq)$	+1.8
$NiO_2(s) / Ni^{2+}(aq)$	+1.7
$Au^+(aq) / Au(s)$	+1.68
$Ce^{4+}(aq) / Ce^{3+}(aq)$	+1.61
$MnO_4^-(aq) / Mn^{2+}(aq)$	+1.51
$Au^{3+}(aq) / Au(s)$	+1.5
$ClO_3^-(aq) /Cl_2(g)$	+1.47
$BrO_3^- / Br^-(aq)$	+1.44
$Cl_2(g) / Cl^-(aq)$	+1.358
$Cr_2O_7^{2-} / Cr^{3+}(aq)$	+1.33
$N_2H_5^+(aq) / NH_4^+(aq)$	+1.24
$MnO_2(s) / Mn^{2+}(aq)$	+1.23
$Pt^{2+}(aq) / Pt(s)$	+1.2
$IO_3^-(aq) / I_2(aq)$	+1.195
$ClO_4^-(aq) / ClO_3^-(aq)$	+1.19
$Br_2(l) / Br^-(aq)$	+1.066
$AuCl_4^- / Au(s)$	+1
$Pb^{2+}(aq) / Pb(s)$	+0.987
$NO_3^-(aq) / NO(g)$	+0.96
$NO_3^-(aq) / HNO_2(aq)$	+0.94
$Hg^{2+}(aq) / Hg_2^{2+}(aq)$	+0.92
$Hg^{2+}(aq) / Hg(l)$	+0.855
$Ag^+(aq) / Ag(s)$	+0.7994
$Hg_2^{2+}(aq) / Hg(l)$	+0.789
$Fe^{3+}(aq) / Fe^{2+}(aq)$	+0.771
$SbCl_6^-(aq) / SbCl_4^-(aq)$	+0.75

$[PtCl_4]^{2-}(aq)\ /\ Pt\ (s)$	+0.73
$O_2\ (g)\ /\ H_2O_2(aq)$	+0.682
$[PtCl_6]^{2-}(aq)\ /\ [PtCl_4]^{2-}(aq)$	+0.68
$H_3AsO_4\ (aq)\ /\ H_3AsO_3\ (aq)$	+0.58
$I_2\ (s)\ /\ I^-(aq)$	+0.535
$TeO_2\ (s)\ /\ Te\ (s)$	+0.529
$Cu^+(aq)\ /\ Cu\ (s)$	+0.521
$[RhCl_6]^{3-}\ (aq)\ /\ Rh\ (s)$	+0.44
$Cu^{2+}(aq)\ /\ Cu\ (s)$	+0.337
$HgCl_2\ (s)\ /\ Hg\ (l)$	+0.27
$AgCl\ (s)\ /\ Ag\ (s)$	+0.222
$SO_4{}^{2-}(aq)\ /\ SO_2\ (g)$	+0.2
$SO_4{}^{2-}(aq)\ /\ H_2SO_3\ (g)$	+0.17
$Cu^{2+}(aq)\ /\ Cu^+(aq)$	+0.153
$Sn^{4+}(aq)\ /\ Sn^{2+}(aq)$	+0.15
$S(s)\ /\ H_2S(aq)$	+0.14
$H^+(aq)\ /\ H_2(g)$ (reference electrode)	0
$N_2O(g)\ /NH_3OH^+(aq)$	-0.05
$Pb^{2+}(aq)\ /\ Pb(s)$	-0.126
$Sn^{2+}\ (aq)\ /\ Sn\ (s)$	-0.14
$Ni^{2+}(aq)\ /\ Ni\ (s)$	-0.25
$Co^{2+}(aq)\ /\ Co\ (s)$	-0.28
$TI^+(aq)\ /\ TI\ (s)$	-0.34
$Se\ (s)\ /\ H_2Se\ (aq)$	-0.4
$Cd^{2+}(aq)\ /\ Cd\ (s)$	-0.403
$Cr^{3+}(aq)\ /\ Cr^{2+}(aq)$	-0.41
$Fe^{2+}(aq)\ /\ Fe\ (s)$	-0.44
$CO_2\ (g)\ /\ (COOH)_2\ (aq)$	-0.49
$Ga^{3+}(aq)\ /\ Ga\ (s)$	-0.53
$Cr^{3+}(aq)\ /\ Cr\ (s)$	-0.74
$Zn^{2+}(aq)\ /\ Zn\ (s)$	-0.763
$H_2O\ (l)\ /\ H_2\ (g)$	-0.8277
$Cr^{2+}(aq)\ /\ Cr\ (s)$	-0.91
$Mn^{2+}(aq)\ /\ Mn\ (s)$	-1.18

V^{2+} (aq) / V (s)	-1.18
Zr^{4+}(aq) / Zr (s)	-1.53
Al^{3+} (aq) / Al (s)	-1.66
H_2 (g) / H^- (aq)	-2.25
Mg^{2+}(aq) / Mg (s)	-2.37
Na^+(aq) / Na (s)	-2.714
Ca^{2+} (aq) / Ca (s)	-2.87
Sr^{2+}(aq) / Sr (s)	-2.89
Ba^{2+}(aq) / Ba (s)	-2.9
Rb^+(aq) / Rb (s)	-2.925
K^+(aq) / K (s)	-2.925
Li^+(aq) / Li (s)	-3.045
OOH^- (aq) / OH^- (aq)	0.88
NH_2HO (aq) / N_2H_4 (aq)	0.74
ClO_3^- (aq) / Cl^- (aq)	0.62
MnO_4^- (aq) / MnO_2(s)	0.588
MnO_4^- (aq) / MnO_4^{2-} (aq)	0.564
NiO_2 (s) / $Ni(OH)_2$ (s)	0.49
O_2(g) / OH^-(aq)	0.4
ClO_4^- (aq) / ClO_3^-(aq)	0.36
Ag_2O (s) / Ag (s)	0.34
NO_2^- (aq) / N_2O(g)	0.15
N_2H_4(aq) / NH_3(aq)	0.1
$[Co(NH_3)_6]^{3+}$(aq) / $[Co(NH_3)_6]^{2+}$(aq)	0.
HgO (s) / Hg (l)	0.0984
NO_3^-(aq) / NO_2^-(aq)	0.01
MnO_2(s) / $Mn(OH)_2$(s)	-0.05
CrO_4^{2-}(aq) / $Cr(OH)_3$(s)	-0.12
$Cu(OH)_2$(s) / Cu(s)	-0.36
$Fe(OH)_3$(s) / $Fe(OH)_2$(s)	-0.56
H_2O / H_2(g)	-0.8277
NO_3^-(aq) / N_2O_4(g)	-0.85
$Fe(OH)_2$(s) / Fe(s)	-0.877
SO_4^{2-}(aq) / SO_3^{2-}(aq)	-0.93

$N_2(g) / N_2H_4(aq)$	-1.15
$[Zn(OH)_4]^{2-}(aq) / Zn(s)$	-1.22
$Zn(OH)_2(s) / Zn(s)$	-1.245
$Cr(OH)_3(s) / Cr(s)$	-1.3
$SiO_3{}^{2-}(aq) / Si(s)$	-1.7

3.2. Electrical relationships

3.2.1. Faraday's law of electrolysis

Michael Faraday (1791–1867) observed that redox reactions occur when electric current passes through electrolyte solutions. When external source supplies electric current to the electrolytic cell, the electrons flow in the cell is in the opposite direction of that in the external circuit. At the cathode, electrons are supplied from the external source, which causes reduction reaction to occur. Similarly, oxidation reaction occurs at the anode since electrons return to the external source from this electrode.

Faraday's law states that the amount of electrochemical reaction occurs at an electrode is proportional to the electric charge Q passing through the cell that is.

$$w = ZQ \tag{8}$$

where w is the product weight, Z is the electrochemical equivalent, the constant of proportionality, Q is the electric charge:

$$Q = It \tag{9}$$

where, I is the current measured by amp (A), and t is the time measured by second (s)

$$w = ZIt \tag{10}$$

One mole of electrons is called one faraday. Faraday constant F is defined as the charge of one mole of electrons. To produce 1 g of reaction product at the electrode, 96,487 C (Faraday constant F) is required, which can be calculated as in Eq. 11:

$$F = N_A e \tag{11}$$

where N_A is Avogadro number, e is the electron charge.

$$1 \text{ Faraday} = 9.64870 \times 10^4 \left(C \text{ mol}^{-1} \right) \tag{12}$$

To produce one mole of hydrogen molecules H_2, two faradays are required since the reduction of $2H^+$ requires two electrons; for one mole of silver, one faraday is required since the reduction of Ag^+ requires one electron; and for one mole of aluminum, three faradays are required since the reduction of Al^{3+} requires three electrons. Generally,

$$w_{eq} = \frac{A_{wt}}{n} \tag{13}$$

where A_{wt} is the atomic weight, n is the number of electrons [32].

If $Q = 1C$, then

$$w_{Q=1} = Z \tag{14}$$

Equation 14 shows that the electrochemical equivalent Z is the weight produced or lost.

Combining Eqs. (8) and (14) produces:

$$w = w_{Q=1}Q \tag{15}$$

Since 96,487 C is required for the deposition of an equivalent of a metal, w_{eq}

$$w_{eq} = 96,487 \, Z \tag{16}$$

And

$$Z = w_{Q=1} = \frac{w_{eq}}{96,487} = \frac{w_{eq}}{F} \tag{17}$$

Using Eq. (13):

$$Z = \frac{A_{wt}}{nF} \tag{18}$$

$$w = ZQ = \frac{A_{wt}}{nF}Q \tag{19}$$

3.2.2. Current efficiency

The current efficiency of metal electrodeposition is highly determined by the deposition rate and energy. For instance, in copper electrodeposition the current efficiency decreases with increasing deposition current density and concentration of acidic solution. However, it is directly proportional with increasing copper concentration, solution flow rate, and solution temperature [34]. Based on Faraday's law, the amount of chemical change at the electrode is proportional to the quantity of electricity passing through the solution. Therefore, the quantity of metal deposited at the cathode or dissolved at the anode is our concern, and the elaborated hydrogen or oxygen at the cathode or the anode, respectively, is considered electrical energy dissipation and process efficiency reduction. Equation (20) is a statement of percentage current efficiency:

$$ CE\% = \frac{Act}{Theo} \times 100 \tag{20} $$

where CE is current efficiency, Act is the actually deposited weight, and Theo is the expected deposited weight. Current efficiency when applied to the cathode reaction is called cathode efficiency and current efficiency as applied to the anode reaction is called anode efficiency [26].

3.2.3. Deposit thickness

Michael Faraday has formulated many laws. But the most important law in this regard is the relationship between current, time, and the weight of the deposit metal, represented by the following equation:

$$ w = (I.t.A)/(n.F) \tag{21} $$

where, w (g) is the weight of deposit metal, I (C/s) is the current, t (s) is the time, A (g.mol^{-1}) is the atomic weight, n (equivalents.mol^{-1}) is the valence of the dissolved metal in solution and F is Faraday's constant (96,485.309 C/equivalent).

Consider nickel electrodeposition at the cathode, nickel ions will be reduced or deposited as nickel metal.

$$ Ni^{2+} + 2e^- \rightarrow Ni \tag{22} $$

If a current of 1 amp is supplied to the cathode for 1 h.

$$ w = \frac{(1 \times 3600 \times 58.69)}{(2 \times 96485.309)} = 1.09 \text{ g} $$

Hence, the thickness is:

$$ T = {}^{w}\!/_{\rho \times a} \tag{23} $$

where T (μm) is the thickness, ρ (gm.cm^{-3}) is the density, and a (cm^2) is the surface area of the deposit.

Combining Eqs., 21 and 23 yields:

$$T = \frac{(I \times t \times A)}{(n \times F \times \rho \times a)} \tag{24}$$

Assume that, the surface area of the Ni deposit is 22.1 cm^2. Thus,

$$T = \frac{(1 \times 3600 \times 58.69 \times 10000)}{(2 \times 96485.309 \times 8.90 \times 22.1)} = 55.67 \ \mu m$$

From the above equation, thickness is proportional inversely with the surface area, provided that, all other parameters in the equation are maintained the same [35].

4. Research and development

Electrodeposited thin films are playing a major role in the development of new materials such as magnetic materials, where electrodeposition reveals strong potential to grow high-quality ultrathin magnetic layers with interesting magnetic properties, and, further, in the study of magnetism on micro- and nano scales [36]. Thermoelectronic devices are another emerging technology that uses electrodeposited materials [37]. J.F.K. Cooper, et al., have successfully grown thin films of ternary alloy of CoNiCu by single-bath electrodeposition techniques using weak and strong electrolytic solutions, whereby the potential can be controlled to grow a wide range of homogeneous films [38]. Nanostructured deposits of the alloys of bismuth, antimony, tellurium, and select other elements have useful thermoelectric properties. One successful fabrication technique is to electrodeposit the materials into nanoscale pores [39]. Electrodeposition is a simple and flexible method for the synthesis of one-dimensional (1D) nanostructures of metals, semiconductors, and polymers in the form of nanowires, nanorods, nanotubes, and nanoribbons that have shown novel physical properties and amazing potential applications in nanodevices [40]. The photovoltaic (PV) conversion of solar energy is one of the most promising renewable energy sources for our future energy needs. Electrodeposition technique was used successfully to synthesize nanostructured electrodes of $ZnxCd_{1-x}$ (O) on ITO glass substrate with several values of x in an aqueous medium containing Zn^{2+} and Cd^{2+} species for energy generating photovoltaic cell, and the efficiency achieved was 1.6% [21]. Electrodeposition of materials has the potential to change phase of deposits upon heating or cooling. Deposit in the amorphous state at room temperature can change phase to crystalline when annealing at elevated temperatures. The switching between amorphous and crystalline phases is useful for rewritable optical storage and solid-state memory [41]. Wiring of printed circuits and interconnections of multilayer microchips are another developed areas of electrodeposition for electronic applications. The plated area can be accurately defined by using photoresist. Electroless plating or Autocatalytics permits a conducting layer to be deposited on insulating materials. For connection, a solder thin layer can be deposited [42].

5. Conclusion

Electrodeposition is the application of metallic coatings to metallic or other conductive surfaces by electrochemical processes for the purpose of good appearance, protection, special surface properties, and engineering or mechanical properties. Electrodeposition is a multidisciplinary technique and based on chemistry, physics, chemical and electrical engineering, metallurgy, and probably others. Besides, it retains the aspects of art, in which experience is the only way to learn. No text on electroplating will produce an expert electroplater. There is no substitute for experience and what is somewhat inelegantly termed know-how. The most beautiful feature of electrodeposition technique is that the deposit thickness, deposition temperature, and deposition potential can be relatively controlled.

Author details

Souad A. M. Al-Bat'hi

Address all correspondence to: su3ad@iium.edu.my

Department of Manufacturing and Materials Engineering, Faculty of Engineering, International Islamic University, Malaysia

References

[1] Gleiter. H., Nanostructured Materials, Basic Concepts and Microstructure, Acta Mater. (1999), 48, 2000, 1-29.

[2] Seigel, R.W., Nanostructured materials-mind over matter, Nanostruct Mater. (2003), 3, 1-6, 1-18.

[3] Seigel, R.W., Nanophase materials assembled from atom clusters, Mater Sci Engin: B. (1993), 19, 1-2, 37-43.

[4] Gleiter, H., State of the art and perspectives, Nanostruct Mater. (2001), 3, 3-14

[5] Wang, Y. and Herron, N., J Phys Chem. (1991), 95, 525.

[6] Witt H. J de, WittmeC. H. M. r, and Dirne F. W. A., Adv Mater. (1991), 3, 356.

[7] Boakye, E., J Coll Interf Sci. (1994), 163, 120.

[8] Aricò A. S., et al., Nanostructured materials for advanced energy conversion and storage devices, Natur Mater. (2005), 4, 366 – 377.

[9] Niederber, M., Nonaqueous sol-gel routes to metal oxide nanoparticle, Acc Chem Res. (2007), 40 (9), 793-800.

[10] Xu, P., Han, X., Wang, C., Zhang, B., Wang, X., and Wang, H. L., Facile synthesis of polyaniline-polypyrrole nanofibers for application in chemical deposition of metal nanoparticles, Macromol Rapid Commun. (2008), 29 (16), 1392-1397.

[11] Chia, H. C., and Yeh, C. S.,Hydrothermal synthesis of SnO_2 nanoparticle and their gas-sensing of alcohol, J Physic Chem C. (2007),111 (20), 7256-7259.

[12] Strobe, R. and Pratsinis, S. E., Direct synthesis of maghemite, magnetite and wustite nanoparticle by flame spray Pyrolysis, Adv Powder Technol. (2009), 20 (2), 190-194.

[13] Kim, S. W., Fujita, S., and Fujita, S., ZnO nanowires with high aspect ratios grown by metal organic chemical vapor deposition using gold nanoparticle, Appl Physics Lett. (2005), 86 (15) 1-3.

[14] Hakuta, Y., Haganuma, T., Sue, K., Adschiri, T., and Arai, K., Continuous production of phosphor YAG: Tb nanoparticle by hydrothermal synthesis in supercritical water, Mater Res Bull. (2003), 38 (7), 1257-1265.

[15] Souad A. M. Al-Bat'hi. Characterization of Electrodeposited Zn (Se, Te) Thin Films/ Polymer (PEO – Chitosan Blend) Junction for Solar Cells Applications. PhD thesis. Malaya University; 2007.

[16] Saeed S., Razieh K., Masoumeh G., and Mohammad K. A. Electrodeposition of copper oxide nanoparticles on precasted carbon nanoparticles film for electrochemical investigation of anti-HIV drug Nevirapine, Electroanalysis. DOI: 10.1002/elan. 201500027, (2015), WILEY-VCH.

[17] , Manuel M., Elvira G., and Elisa V., Electrochemical synthesis of mesoporous CoPt nanowires for methanol oxidation, Nanomaterials. (2014), 4(2), 189-202.

[18] Gleiter, H., Schimmel, Th., and Hahn, H., Nanostructured solids from nano-glasses to quantum transistors materials assembled from atom clusters, Nano Today. (2014) 9, 17-68.

[19] Xianwen, et al., Nanocarbon-based electrochemical systems for sensing, electrocatalysis, and energy storage, Nano Today. (2014) 9, 405-432.

[20] Eeu, Y.C., Lim, H.N., Lim,Y.S., Zakarya, S.A., and Huang, N.M., Electrodeposition of polypyrrole/reduced graphene oxide/iron oxide nanocomposite as supercapacitor electrode material, J Nanomater. (2013), Volume 2013, Article ID 653890 Hindawi Publishing Corporation.

[21] Umer M., Souad. A. M. Al-Bat'hi, Zn_xCd_{1-x} (O) thin film nanorods for PV applications, Int J Nano Sci Technol. (2013),1, (3), April. 09-16.

[22] Yu-Huei H. and Yu-Chen T., Electrodeposition of platinum and ruthenium nanoparticles in multiwalled carbon nanotube-nafion nanocomposite for methanol electro oxidation, J Nanomater. Volume 2009, (2009)

[23] Guangwei S., Lixuan M., and Wensheng S., Electrodeposition of one-dimensional nanostructures, Recent Patents Nanotechnol. (2009), 3, 182-191.

[24] Zein El Abedin, S., Pölleth, M., Meiss, S.A., Janek, J., and Endres, F., Ionic liquids as green electrolytes for the electrodeposition of nanomaterials, Green Chem. (2007), 9, 549-553.

[25] http://www.gitam.edu/eresource/nano/nanotechnology/

[26] http://nzic.org.nz/ChemProcesses/metals/8G.pdf

[27] Chemistry.tutorvista.com

[28] http://www.gcsescience.com

[29] http://chemistry.about.com/od/chemistryglossary/g/weak-electrolyte-definition.htm

[30] http://chemistry.about.com/od/waterchemistry/tp/Strong-and-Weak-Electrolytes.htm

[31] http://www.physics.usyd.edu.au/super/life_sciences/E/E5.pdf

[32] Milan P., Mordechay S., and Dexter D., Modern Electroplating, Fifth Edition (2010), John Wiley & Sons, Inc.

[33] www.chemeddl.org/services/moodle/media

[34] Pavlovi M. G., et al., The current efficiency during the cathodic period of reversing current in copper powder deposition and the overall current efficiency, *J Serb Chem Soc.* (2003) 68 (8–9) 649-656. JSCS – 3073.

[35] http://www.electrolytics.org/faradaysLaw.html

[36] Allongue P., et al., *Surf Sci.* (2009), 603, 1831.

[37] Xiao F.,et al., Electrochem *Acta.* (2008), 53, 8103.

[38] Cooper, J.F.K., et al., The effects of electrolyte concentration on film composition and homogeneity in electrodeposition, Electrochem Commun. (2013) 27, 96-99.

[39] Kouiharenko, E., et al., J Micromech Microengin. (2008), 18,104015.

[40] RemSkar, M., Inorganic nanotubes. Adv Mater. (2004), 16, 1497-1504.

[41] Huang, Q., et al., J Electrochem Soc. (2008), 155 (2), DI04.

[42] Datta, M. and Landolt, D. Electrochem Acta. (2000), 45, 5235.

3

Aluminum Anodic Oxide AAO as a Template for Formation of Metal Nanostructures

Piotr Tomassi and Zofia Buczko

Abstract

The aim of the chapter is to describe the applications of AAO as a template in metal nanostructures formation and to present the experimental results obtained by authors in this field. The basic mechanism of the process of anodic oxidation of aluminum was described. The influence of oxidation parameters on the AAO structure was discussed. The processes of electrochemical metal deposition in AAO were described. The main present as well as future applications of metal nanostructures formed were listed.

Keywords: anodic oxidation of aluminum (AAO), metal nanostructures

1. Introduction

The aim of this chapter is to describe the applications of AAO as a template in metal nanostructures formation and to present the experimental results obtained by authors in this field.

Metals in a state of high dispersion currently play an important role in technology. Their chemical and physical macroscopic properties, such as the rate of their reaction with other substances, colours and mechanical properties are significantly different from the bulk metals. Fundamental research and technology development over the last decade have resulted in wider use and implementation of metal-containing materials in a state of high dispersion into the industrial practice.

Solid nanoparticles are a system of unstable thermodynamic state, so even precious metals are not found in this form in nature. However, in the case of metals having a high cohesive energy, and thus a high melting point, it is possible to obtain and maintain them in the form of nanodispersion for a long period of time. This is the case of high activation energy of the agglomeration process of particles, which is a crucial parameter. This applies in particular to transition metals. Nanodispersions of these metals are most often applied in practice.

The basic properties of highly dispersed metals are described in monographs of Romanowski [1] and also of Feldheim and Foss [2]. The authors point out that the systems containing metal nanoparticles are particularly interesting because their synthesis and chemical modification are simple. The main directions, according to the authors, of current basic research are given in the following text.

1.1. Investigation of optical properties

Metal particles often exhibit strong plasmon resonance extinction bands in the visible spectrum. While the spectra of molecules can be understood only in terms of quantum mechanics, the plasmon resonance bands of nanoscopic metal particles can often be described in terms of classical free-electron theory and electrostatic models for particle polarizability. In contrast to molecular systems, the linear optical properties of metal nanoparticle composites can be changed significantly without a change in essential chemical composition.

Optical properties of metal nanoparticles are usually described by Mia theory. Material testing is mostly done by optical spectroscopy. Deviation from the theoretical and measured values is unavoidable due to uneven sizes and shapes of metal particles in real systems and the diversity of their distribution in a material. Therefore, theoretical analytical methods are not enough to describe the real particle systems and numerical methods are often implemented. Investigation results of extinction, absorption, scattering, and optical interaction of metal nanoparticles can be found, among others, in the work of Lin et al. [3] and Noguez et al. [4].

1.2. Investigation of electrical properties

Metal particles and bulk materials are similar taking into account their electrical properties. Simple classical charging expressions and RC equivalent circuit diagrams can be used in both cases to describe surface charging and electron transport processes. Description of electric properties of metal nanoparticles requires knowledge of their size and dielectric properties of the surrounding medium.

Examples of research in this field are given in the work of Wu et al. [5] on study of metal clusters of silver, gold, and copper. Static electric polarizability and absorption spectrum have been measured. Density functional theory (DFT) has been implemented for description of investigated systems. A competition of charge transfer and electron cloud distortion was proposed for explaining the spacing dependence of both electric and optical properties.

1.3. Investigation of magnetic properties

Small, single-domain ferromagnetic particles, containing a few tens or a few hundred atoms have a large total magnetic moment. When these ferromagnetic particles are separated by comparatively large distances, their magnetic moments do not interact strongly and they form a system of magnetic dipoles, whose relative orientation can be randomized by thermal motion even in relatively low temperatures. The overall magnetization of the system is therefore similar to that of a paramagnetic material, expressed by the Langevin equation.

The study of ferromagnetic nanoparticles systems are mainly targeted to the specific use of new nanomaterials. In particular, researches are provided to receive improved materials for high-density data storage, for the construction of high-speed computing components, and for a new-generation high-performance display.

The test results of the magnetic properties of metal nanoparticles have been published, among others, by Kim et al. [6]. The authors investigated Ni-Co, Ni-Fe, and Co-Pt alloy nanoparticles. Magnetic hysteresis loops were presented and values of magnetic susceptibility, the saturation magnetization, and coercivity were calculated.

Implementation of porous alumina, obtained by anodic oxidation of aluminum, as a template for producing new nanostructured materials is a quickly developing part of nanotechnology in recent years. Introducing a metal by electrodeposition inside the AAO pores, it is possible to obtain a durable material, characterized by a uniform structure of nanoparticles of a single metal or alloy in a matrix of dielectric amorphous alumina. The basis of manufacturing processes of composite materials of this type is presented in the following sections. References are listed in Table 1.

No.	Subject	References
1.	Basic properties of highly dispersed metals	[1, 2]
2.	Optical properties of metal particles	[3, 4]
3.	Electrical properties of metal particles	[5]
4.	Magnetic properties of metal particles	[6]
5.	The theories of anodic oxidation of aluminum	[7–12]
6.	Theoretical modeling of porous oxide growth on aluminum	[13–19]
7.	Preparation of AAO as a template in nanotechnology	[20–48]
8.	The processes of metal electrodeposition on AAO template	[49–68]

Table 1. Basic studies on AAO as a template for formation of metal nanostructures

2. Anodic oxidation of aluminum

Passivation phenomenon of aluminum spontaneously occurs in contact with oxygen and thin film of alumina on the surface of the metal is formed. Thicker oxide coatings can be produced by the use of electrochemical anodic polarization in selected solutions. This surface finishing of aluminum work pieces is widely applied for corrosion and abrasion protection and for decorative purposes.

In carrying out the anodizing process of aluminum in an electrolyte, which slightly dissolves alumina, an oxide layer is formed with a unique porous structure on the metal surface. Highly ordered pores are perpendicular to the base and pass through almost the entire thickness of the oxide layer. Directly on the metal surface nonporous barrier layer is grown and it usually has a thickness of several tens of nanometers. When an aluminum substrate with high purity and uniform structure is used and the parameters of the operation of the electrochemical oxidation are carefully controlled, homogeneous distribution of pores as a hexagonal network can be obtained. The pore diameters are uniform; it depends on the oxidation conditions and can be controlled in the range from 10 to several hundred nanometers. The total thickness of the oxide layer is determined by the oxidation time. It is usually in the range of tenths of a micron to tens of microns. The schematic drawing of the AAO structure is presented in Figure 1.

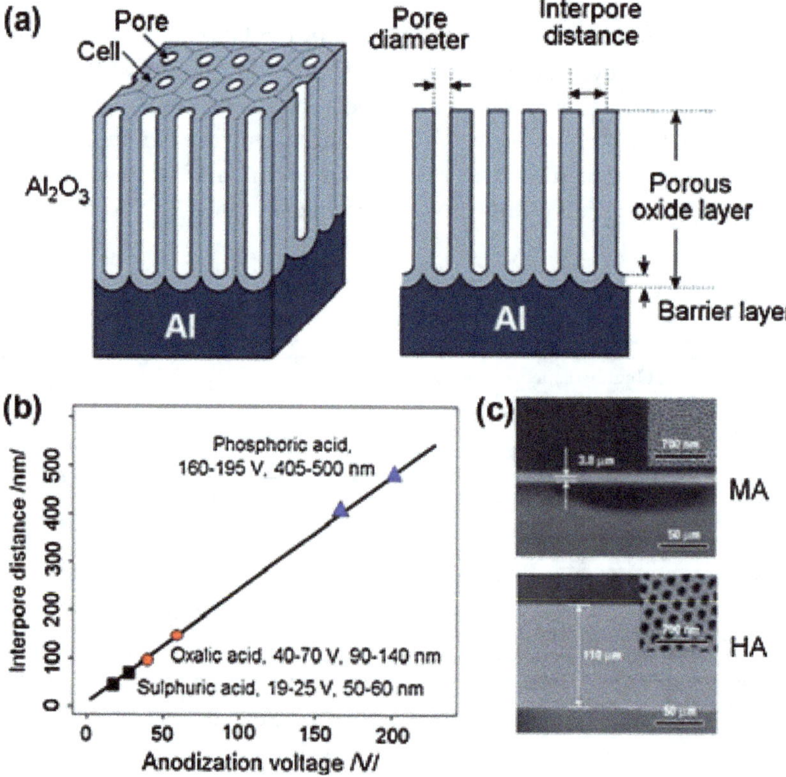

Figure 1. (a) Schematic drawing of AAO structure prepared by electrochemical oxidation of Al. (b) Summary of self-ordering voltage and corresponding interpore distance of AAO produced within three well-known regimes of electrolytes (sulfuric, oxalic, and phosphoric). (c) (Top) SEM cross-sectional view of AAO membrane formed by MA (0.3 M $H_2C_2O_4$, 1°C, 40 V) and (bottom) by HA (at 140 V) for 2 h (insets: SEM top view of pore structures). Permission Elsevier [61].

The mechanism of the process of anodic oxidation of aluminum and forming the porous oxide layer is still under investigation. There are some basic theories, like the Keller, Hunter, and Robinson geometric model [7] and the colloidal one of Murphy and Michelson [8]. It is believed that the porous structure is a result of two competitive reactions: Al_2O_3 forming and dissolution of the oxide. These reactions are stimulated by the electric field that is distributed inside the barrier layer. The electric field intensity inside the barrier layer is of the order of 1 V/nm during the oxidation process. Such high field is needed to excite the ionic current inside the oxide. This field also stimulates the dissolution reaction ("field-assisted dissolution"). Also, due to the tunneling effect, additional electron current may occur. The role of that current in the anodic oxidation of aluminum was described by Palibroda [9-10].

A summary of recent interpretations of the reaction mechanism of anodic oxidation of aluminum can be found in the Brace monograph [11] and the Wielage et al. work [12]. The total chemical reaction is as follows:

$$2Al + 3H_2O \rightarrow Al_2O_3 + 6H^+ + 6e^- \tag{1}$$

It consists of two partial reactions:

$$2Al \rightarrow 2Al^{3+} + 6e^- \tag{2}$$

$$3H_2O \rightarrow 6H^+ + 3O^{2-} \tag{3}$$

The oxide forms due to migration of Al^{3+} ions from basic metal into the solution, while the movement of ions O^{2-} is in the opposite direction. Under the influence of a high electric field, there is ionic conductivity in the oxide layer where the aluminum and oxygen ions are charge carriers.

A characteristic porous structure of the alumina layer is formed as a result of the chemical and electrochemical dissolution of the oxide. Dissolution reaction is promoted by local increase in hydrogen ion concentration (reaction 1) and high electric field inside the barrier layer ("field-assisted dissolution"). Due to the high electrical resistance of the barrier layer, the Joule heat is given off during a flow of electric charge, which causes local increase in temperature and also enhances the dissolution reaction of the oxide.

Defects and impurities, which are always present in the metal substrate, are precursors of pores. Due to the contact with an aqueous solution to the oxide layer surface, there are always various intermediate forms of hydrated oxide, including aluminum hydroxide. In addition, due to the adsorption of anions from the solution, the oxide material is enriched with other compounds, like for example sulfates in the case of processing in sulfuric acid.

Oxide layer with porous structure is obtained using solutions of dibasic and tribasic acids, in which the oxide is dissolved. When the critical thickness of the barrier layer is obtained, pores are initiated in flows, cracks, and impurities spots. Only some part of embryo pores develops up to form final porous structure with the hexagonal arrangement. It is determined by the reactions' activation energy of oxide formation and dissolution, also by the distribution of electric field in the barrier layer

Attempts have been made to create modeling and mathematical description of the electro-chemical processes of formation of porous aluminum oxide layers [13-18]. Parkhutik and Shershulsky [13] are the first to prove that the potential distribution inside the oxide layer can be calculated with the use of Laplace equation. Calculations with the adequate assumptions of the local moving rate of the solution–oxide and oxide–metal phase boundaries have been made for a number of cases in the two-dimensional system. As the results pictures of the development of porous structures have been obtained similar to those observed on the SEM images of experimental samples. Also, numerical approximations are often used for the mathematical modeling [19].

Beside the above-described mechanism of aluminum anodic oxidation process and its mathematical modeling, the authors point to additional important factors that have an impact on the process of forming a porous oxide structure:

1. The volume of the oxide is higher than the metal consumed and therefore there are strong strains in oxide layer that cause mechanical stress fractures (cracks). These cracks are often the beginning of a pore.

2. The process of oxidation may be affected by local differences in the wettability of the surface of the oxide.

3. Part of the aluminum ions Al^{3+} is ejected from the metal to the solution without binding to the oxide structure. This phenomenon reduces the current efficiency of anodic oxidation reaction.

4. In the presence of phases containing foreign elements in aluminum substrate, it is possible, as polarization is anodic, side reaction of evolution of oxygen gas to occur. Like the previous phenomenon, it reduces the current efficiency of the oxidation reaction.

5. Barrier layer thickness, the distance between the pores and their diameters are propor-tional to the applied voltage, with the other process parameters being fixed.

On the basis of these assumptions, Wu et al. [17] attempted to demonstrate that the chemical and electrochemical reactions of oxide dissolution do not have significant effect on the process of forming a porous oxide structure. This statement, however, is not consistent with the conclusions of other authors.

3. Preparation of AAO template

A porous oxide coating can be obtained by anodic oxidation of aluminum using a number of different types of electrolyte solutions. Sulfuric, oxalic, phosphoric, or chromic acids solutions are typically implemented. The porous anodic layer can also be obtained from solutions of many organic acids, especially the polybasic acids, such as tartaric, citric, sulfosalicylic, maleic, and succinic acids.

Such coating parameters as the thickness of the barrier layer, pore diameter, the distance between the pores and their surface density considerably vary depending on the type of processing solutions, as is shown in Tables 2 and 3 [20]. For a given type of solution, the geometric parameters of the net of pores can be controlled by varying the voltage during anodic oxidation. The thickness of the barrier layer and the diameter of the pores are proportional to the applied voltage (Table 2). The pore diameter is usually from 10 to a few tens of nanometers. Pores having higher values of diameter can be obtained from solutions of phosphoric acid or organic acids at high voltages. Li et al. [21] give the equation that correlates distance between the pores and the voltage and pH value for processing solutions. Uniform grid of unusually high pore diameters of up to 200–500 nm have been successfully obtained with properly selected oxidation conditions in solutions containing phosphoric acid and/or organic acids with high voltage current in [22-26]. In the solution of citric acid using high current voltage, Mozalev et al. [27] obtained the interpore distance 1.1 µm, barrier layer thickness 0.5 µm, and pore diameter 0.23 µm. Also, lowering the temperature of anodic oxidation, thicker coatings can be easier to obtain; their porosity is lower and hardness is higher. The examples of SEM images of AAO coatings are presented in Figures 2 and 3.

Electrolyte	Conc. % (wt)	Temp. °C	Barrier layer thickness nm/volt	Pore diameter nm/volt	Wall thickness nm
Phosphoric acid	4	25	1.19	1.10	33
Oxalic acid	2	25	1.18	0.97	17
Chromic acid	3	40	1.25	1.09	24
Sulfuric acid	15	10	1.0	0.80	12

Table 2. Barrier characteristics for various electrolytes [20]

Electrolyte	Conc. % (wt)	Temp. °C	Voltage, V	Pores per cm^2
Sulfuric acid	15	10	15	76×10^9
			20	52×10^9
			30	28×10^9
Oxalic acid	2	25	20	35×10^9
			40	11×10^9
			60	6×10^9
Chromic acid	3	50	20	22×10^9
			40	8×10^9
			60	4×10^9
Phosphoric acid	4	25	20	19×10^9
			40	8×10^9
			60	4×10^9

Table 3. Pore density in oxide coatings [20]

Figure 2. SEM image of AAO coating obtained in solution of sulfosalicylic acid 50 g/dm^3, oxalic acid 10 g/dm^3, and sulfuric acid 5 g/dm^3, 3 A/dm^2, 20°C, 30 min. Permission Librant [114].

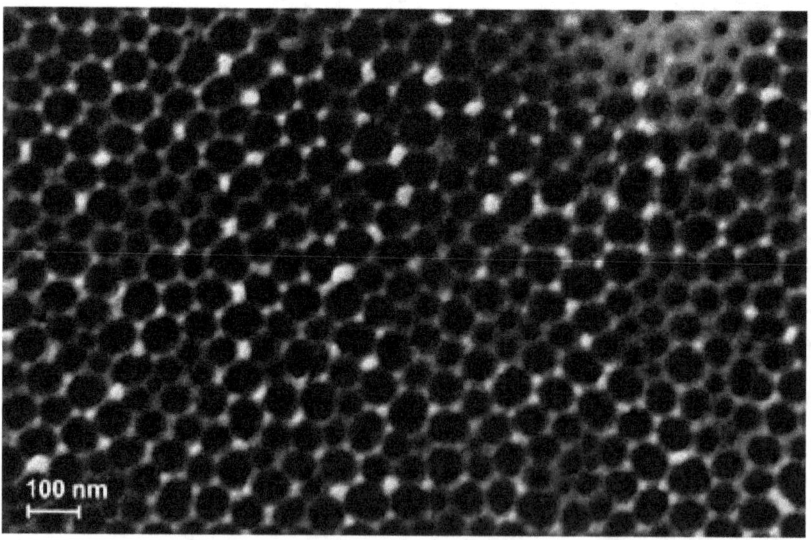

Figure 3. SEM image of AAO coating obtained in solution of sulfosalicylic acid 50 g/dm^3, oxalic acid 10 g/dm^3, and sulfuric acid 5 g/dm^3, 3 A/dm^2, 20°C, 30 min. Permission Librant [114].

For many applications of Al_2O_3 coatings in nanotechnology, including membranes for nanofiltration, it is necessary to separate oxide layers from aluminum substrates. The following techniques are most frequently used:

1. Screening of a part of the electrode surface with lacquer. After anodizing, the lacquer coating is removed, and then the aluminum substrate is entirely dissolved using a solution that is not aggressive to the oxide, e.g., copper chloride solution containing hydrochloric acid [28], mercury chloride solution [29], or a saturated solution of iodine in methanol [30].

2. Programmable voltage reduction in the final stage of anodic oxidation. A thin oxide layer of high porosity and small pore wall thickness is created bottom-up from the substrate. The thickness of the nonporous barrier layer also tends to decrease. As a result, the mechanical properties of the bottom oxide layer are reduced and it can easily be separated from the substrate [31], sometimes in an additional operation of chemical or electrochemical dissolution. For this purpose, for example, Zhao et al. [32] used a cathodic polarization in a solution of potassium chloride.

For applications in nanotechnology, it is very important to obtain the pore distribution, which is homogeneous and uniform over the entire surface in the oxide layer. In order to obtain the highest possible uniformity of the pore distribution, the following techniques are applied:

1. The use of additives to the solutions usually in the form of aliphatic alcohols, glycerol, or ethylene glycol [22,32-35]. The addition of ethanol or methanol facilitates heat removal from the barrier layer during the process of oxidation and reduces the risk of defects in the oxide coating. It also allows the use of high current density, which significantly shortens the creation of oxide coating. Li et al. [36] achieved current density of 4000 A/m^2 during anodic oxidation process in solution phosphoric acid–ethanol.

2. Application of programmed pulse current instead of direct current [37-40]. Anodic and cathodic pulses are used alternately or only anode pulses with a break. According to the authors, the use of pulsed current allows for better structure uniformity of the oxide layer and reduces the risk of defects caused by heat generated in the barrier layer during the oxidation process.

3. The use of a two-step anodizing process. Today it is a commonly used technique which allows increasing the distribution uniformity of the pores and reduces the scatter of their geometrical parameters. Polished aluminum surface is anodically preoxidized. An oxide layer is selectively removed in a subsequent operation. Usually for this purpose the etching solution of phosphoric and chromic acids is used, which does not affect the aluminum substrate. After this operation, the aluminum surface has a scalloped structure, so the homogeneity of the oxide layer formed in the second stage of anodic oxidation increases [41-44].

4. The initial formation of the aluminum surface by mechanical, laser, or other method [45-47]. Zaraska et al. [45] listed the following surface-shaping techniques:

    • Prepatterning using a tip of the scanning probe microscope (SPM) or atomic force microscope (AFM)

    • Focused ion beam lithography

    • Holographic lithography

- Using stamps (molds) with regular array of convexes prepared lithographically

- Optical diffraction grating

- Nanosphere lithography (NSL)

Asoh et al. [46] obtained square layout of cells with centrally located pores after the initial shaping of a surface using the imprinting process. This square system is fundamentally different from the natural hexagonal distribution of pores.

Choi et al. [47] implemented a similar technique and got a uniform distribution of pores having a triangular cross section.

As can be seen there is a great range of possibilities of appropriate selection techniques for producing the porous aluminum oxide layer in order to achieve a certain distribution and pore geometry for use in nanotechnology. For example, a typical technique used to obtain a template for the production of metallic nanostructures is shown in Table 4 [29].

Various tools can be used for evaluation of the uniformity of pore distribution, e.g., Voronoi diagrams, radial distribution function (RDF) [48], fast Fourier transform (FFT) images, Delaunay triangulations (defects map), pair distribution functions (PDF), or angular distribution functions (ADF) [44].

Step no.	Process
1.	Degreasing of AA 1050 alloy in ethanol and acetone
2.	Electrochemical polishing Perchloric acid (60 wt.%) and ethanol (1:4 vol) Constant potential 20 V for 1 min at 10°C
3.	Anodic oxidation in 0.3M oxalic acid Constant potential 45V for 60 min at 20°C
4.	Alumina layer removal by chemical dissolution 6 wt.% H_3PO_4 + 1.8 wt. % $H_2Cr_2O_4$ Time 12 h, temperature 45°C
5.	Anodic oxidation in 0.3M oxalic acid Constant potential 45V for 1, 2, 4, or 8 h at 20°C The samples with different coating thicknesses were obtained
6.	Removal of aluminum substrate by chemical dissolution Saturated $HgCl_2$ solution
7.	Chemical etching of alumina barrier layer 5 wt.% H_3PO_4 at 25°C

Table 4. The example of procedure of AAO template formation [29]

4. Metal deposition

The composite systems AAO–metal can be formed by inexpensive and simple method of electrochemical metal deposition. It does not require complex equipment as in the case of CVD or PVD methods.

The first attempts of metals electrodeposition inside the pores of the anodic oxide layer on aluminum AAO were associated with the development of electrochemical coloring technology. Metal nanoparticles can be deposited into the pores of Al_2O_3 layer by transferring a freshly prepared sample to the solution of the salt of easily reducible metal and then use of cathodic polarization. This causes color effect as a result of optical phenomena like absorption and scattering of light by the deposited metal particles inside the pores. In this way, the oxide layer which was initially colorless can be converted to durably colored; usually there are shades of brown or black.

Electrochemical technologies of coloring anodized aluminum were developed in the 1970s and it is associated with the names of researchers such as Caboni, Langbein, Pfanhauser, Assad, and Sheasby [49]. Herrmann [50] gives 93 examples of electrochemical coloring processes using solutions of salts of the following metals: Pb, Cd, Cr, Fe, Au, Co, Cu, Mn, Ni, Se, Ag, Te, Zn, Sn. If an additional modification of the shape of the pores is introduced before the metal deposition process, it is possible to get a new color effect due to the interference effect [51].

Electrochemical coloring method of anodized aluminum is currently widely used as durable, decorative surface treatment for aluminum components, particularly for construction and architecture applications. Usually, tin nanoparticles are deposited in the pores of AAO using alternating AC or pulse current. When only cathodic polarization is implemented in that process, there is a risk of damage to the oxide layer.

The mechanism of electrodeposition of metals in the pores of the oxide layer was investigated by Skominas et al. [52] and Zemanova et al. [53] and lately by Bograchev et al. [54]. The effects of the reduction reaction of hydrogen ions and of the current frequency were studied.

Wider research in this field was provided by Tomassi [55].

Anodic oxidation of 99.5% Al samples was conducted in 1.75 M H_2SO_4 solution in temperature 20°C with voltage 16 V. Thickness of oxide layer obtained was 15 μm. The oxide layers from phosphoric acid solutions were also obtained. The electrodeposition of nickel, copper, tin, or silver was performed in temperature 25°C in the following solutions:

$$\text{a)}NiSO_4\,0.1\text{ M} \quad \text{b)}CuSO_4\,0.1\text{ M} \quad \text{c)}SnSO_4\,0.1\text{ M} \quad \text{d)}AgNO_3\,0.01\text{ M}$$
$$\text{H}_3BO_3\,0.5\text{ M} \quad\quad H_2SO_4\,0.05\text{ M} \quad\; H_2SO_4\,0.2\text{ M} \quad\quad\; H_2SO_4\,0.1\text{ M}$$

The maximum voltage of alternating current was 16 V (Ni) or 14 V (Cu, Sn, Ag). Usually, the frequency 50 Hz was used, but wide range from 0.1 to 1000 Hz was checked.

Electrochemical investigations have been performed using transient curves method, polarization methods, and impedance spectroscopy.

The uniform composite layers were obtained in the frequency region 10 – 100 Hz. With frequencies below 10 Hz the defects in oxide layer appear. In the high frequency region under 500 Hz the electrodeposition does not occur and high capacity current is observed.

Chemical analyses of obtained composite layers as well as structure, magnetic, microbiological properties, gravimetric and electrochemical investigations have been performed.

The X-ray analysis has confirmed that the metal deposition starts at the bottom of the pores close to the aluminum surface and the volume occupied by metal increases.

The investigations performed by transmission microscopy have shown that all pores are filled by metal deposit (Figure 4). The diameter of metal particles depends on pore diameter and has been found to be from 5 to 30 nm.

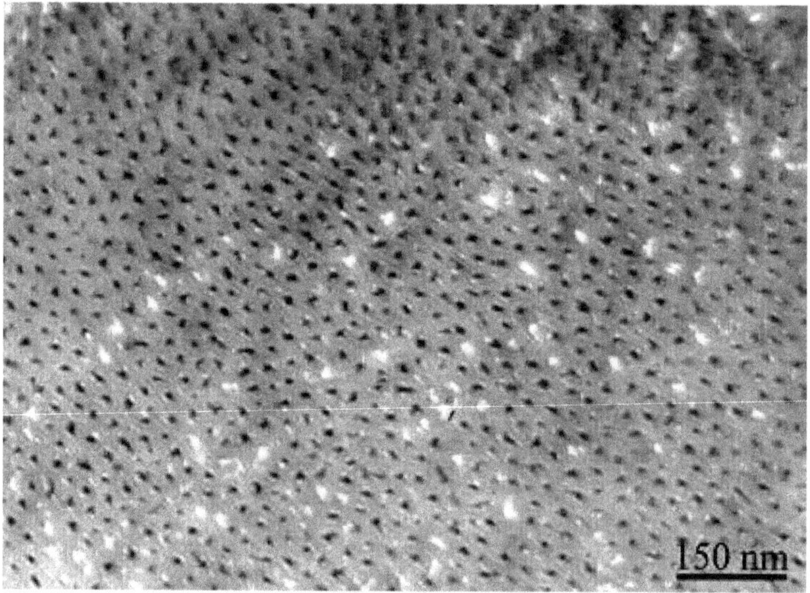

Figure 4. TEM image of AAO obtained in 1.75 M sulfuric acid, 20°C, 16 V, 30 min with nickel nanoparticles, AC nickel deposition, 2 min.

The role of alternating polarization in metal deposition process can be clarified as a result of the investigations. In the cathodic cycles, the hydrogen ions' reduction occurs simultaneously to metal electrodeposition. The diffusion of hydrogen ions inside the oxide layer leads to formation of hydrogen bonds and hydrated regions of higher conductivity (active sites) are created. In these regions, the probability of electron transport is higher than in the other sites, where the current is mainly of ionic character. In the cathodic cycles, the new active sites are formed and filled by metal deposit.

In the anodic cycles, the repassivation of active sites occurs. This process ensures that the barrier layer is not damaged. The metal electrodeposition can then be effectively conducted with the formation of a uniform oxide–metal composite coating.

The optimum current frequencies are in good correlation with a time of charge carriers formation determined for aluminum oxide layers by Ebling [56], Hassel, and others [57-58]. Authors of this chapter have obtained composite layers consisting of the following metals: Ni, Cu, Sn, Fe, Co, Zn, Cd, Au, Pd, Ag.

Metal nanoparticles deposited with the method described above may create durable system with an oxide matrix characterized with interesting scientific and operation properties. Metal nanoparticles can also be separated from the matrix, after, for example, a selective dissolution of alumina, and then examined individually. Nanowires, nanotubes, or nanodots can be similarly obtained. The TEM image of AAO layer with nickel nanoparticles is shown in Figure 4. The bigger diameter of metal nanostructures can be obtained when the oxide matrix is formed in phosphoric acid solution. The AAO formed in phosphoric acid is presented alone in Figure 5 and with nickel nanostructures in Figures 6 and 7.

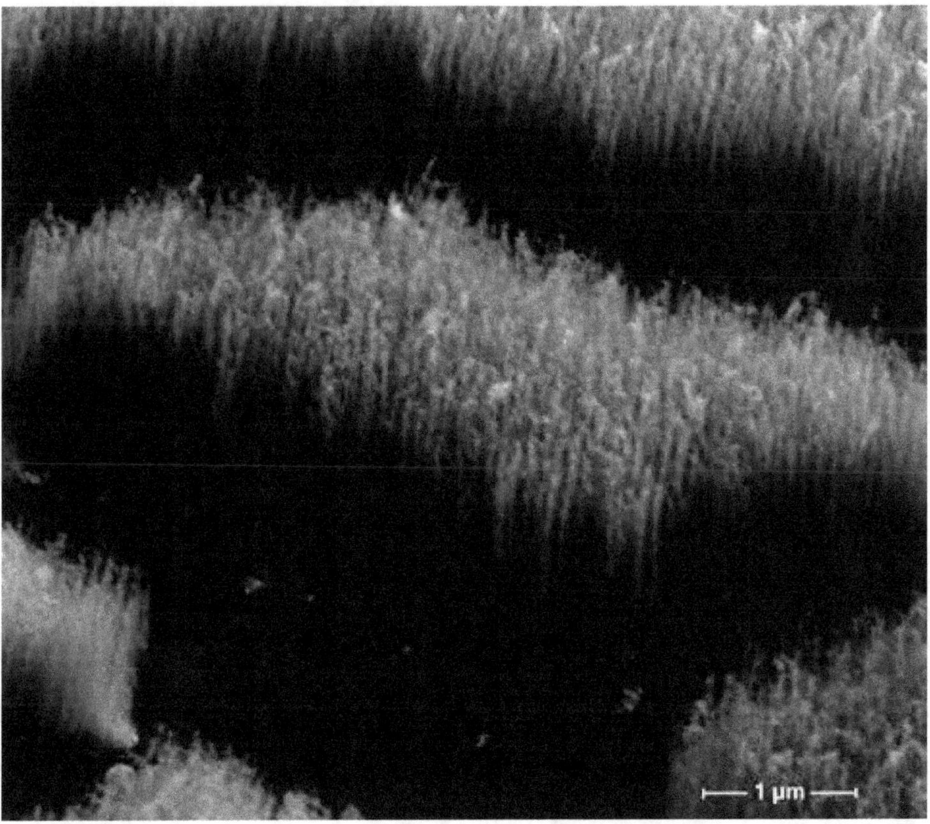

Figure 5. SEM image of AAO obtained in 500 g/dm² phosphoric acid, 30°C, 30 V, 20 min.

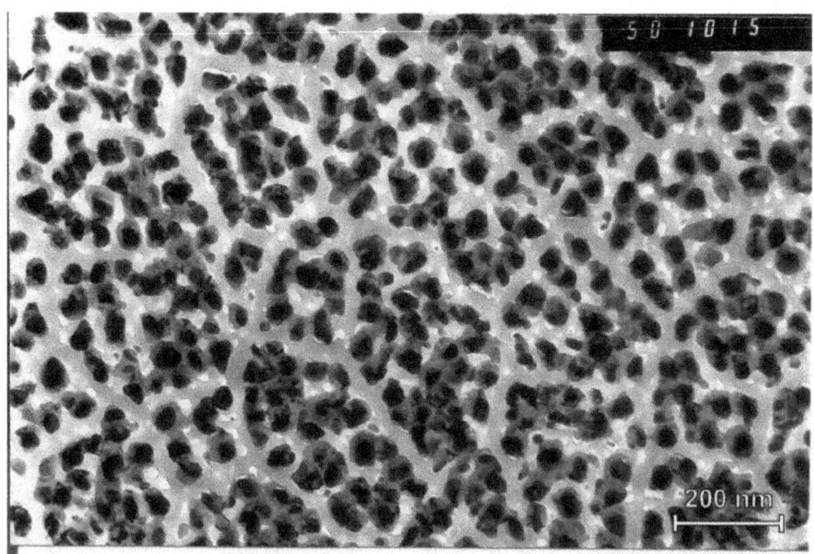

Figure 6. SEM image of AAO obtained in phosphoric acid with nickel particles. AC nickel deposition, 30 V, 30 s.

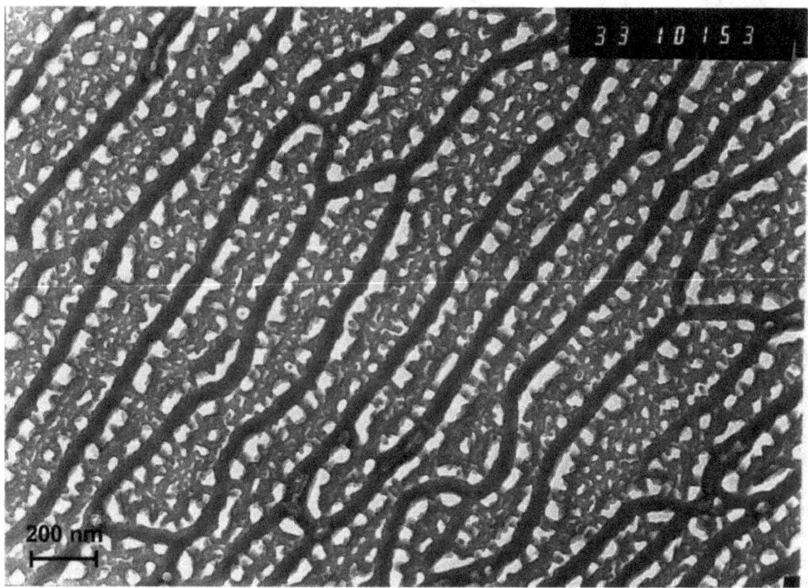

Figure 7. SEM image of AAO obtained in phosphoric acid with nickel particles. AC nickel deposition, 40 V, 15 s.

Oxide layers can be of different porosity with branches when anodic oxidation process is unsteady. The pore branching is clearly visible in the SEM photograph of the oxide layer obtained in a solution of sulfosalicylic acid (Figure 8). After deposition of the metal within the pores of oxide layer, the structure of metal nanoparticles is complex, e.g., in the form of branches [59-60] as is shown in Figure 9. Anodic oxidation was performed in oxalic acid solution; various barrier layer thinning procedures (BLT) were used before metal deposition. Metal structures deposited with DC are usually compact, but when AC pulse current is implemented, they may be of grained structure.

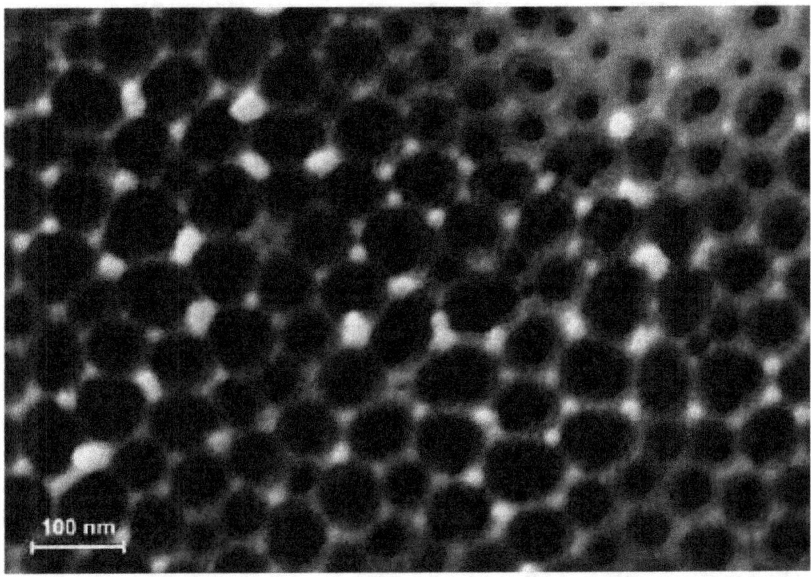

Figure 8. SEM image of AAO coating obtained in solution of sulfosalicylic acid 50 g/dm^3, oxalic acid 10 g/dm^3 and sulfuric acid 5 g/dm^3, 3 A/dm^2, 20°C, 30 min [114].

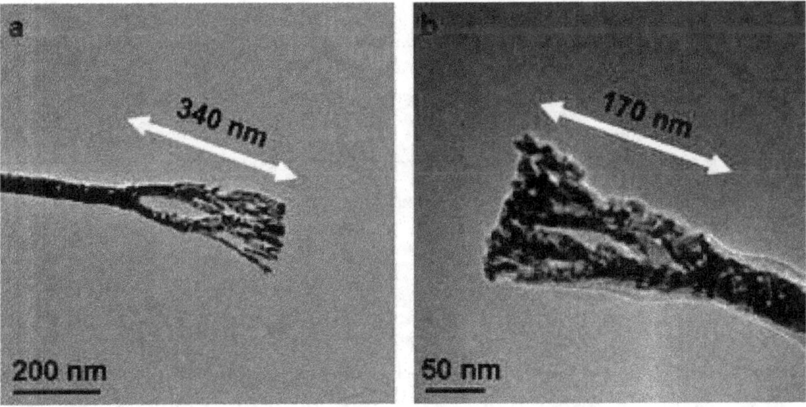

Figure 9. Nickel branches obtained by anodic oxidation in oxalic acid, 20°C, 45 V, 1 h, and nickel pulse electrodeposition from Watt's solution with various BLT procedures (Permission Elsevier [59]).

In the processes for preparing metallic nanostructures with the use of AAO layers, sometimes additional operations are applied in order to obtain geometrically complex systems, e.g., PVD methods, photochemical techniques, and other methods. An excellent overview of these methods can be found in the monograph of Jani et al. [61]. Some examples of complex manufacturing of anodic oxide layer and forming of the metal nanostructures are given in Table 5.

No.	Process	References
1.	Evaporation of aluminum 0.3–1 μm layer on oxide surface. After using a lithographic process the regular patterns of porous oxide were obtained.	[62]
2.	Micropatterned substrates were prepared by fabrication of polyethylene glycol hydrogel microstructures on alumina membranes with 200 nm nanopores using photolitography.	[63]
3.	Electroless deposition of nickel–boron nanoparticle structures on AAO membrane.	[64]
4.	AAO membranes were immersed in $AgNO_3$ solution. As a result of hydrothermal reaction, the Ag nanoflakes were obtained on the oxide membrane surface.	[65]
5.	The AAO membranes were fabricated by a two-step anodizing of AA 1050 alloy followed by removal of Al and pore opening/widening procedure. Au conducting layer was sputtered on one side of AAO membrane. The dense arrays of Ag, Au, and Sn nanowires were fabricated by a DC electrochemical deposition process.	[66]
6.	Thin Au films with highly ordered arrays of hemispherical dots were fabricated by evaporating Au on the surface of porous anodic alumina template.	[67]
7.	Highly ordered arrays of Ni nanoholes and FeNi dots were prepared by sputtering and replica processing techniques using nanoporous alumina membranes as a template.	[68]

Table 5. Examples of complex methods of preparation of oxide templates and metal nanostructures

5. Emerging applications

Electrochemically obtained porous layers of aluminum oxide are widely used in nanotechnology, not only in scientific research but also in various applications. Porous oxide layer can be separated from the substrate. After removing the barrier layer membranes are obtained. They can be applied for example in nanofiltration for removing bacteria from medical preparations (cold sterilization). They are also used to prepare the solutions of test samples for liquid chromatography. Revised filtration properties of these membranes can be found in Lee et al. [69].

Alumina is insoluble, nontoxic, and characterizes excellent biocompatibility. AAO porous layer has a highly developed surface area – about 2 to 3 orders of magnitude higher than the surface of the aluminum substrate. Hence, importance of these materials is growing and they are more widely used in chemical catalysis, in bioreactors, and in sensors. Porous anodic alumina is also an excellent substrate for culturing bacteria in discrete growth compartments. These cultures are investigated with the use of fluorescence microscopy. A broad overview of the use of the porous AAO in biotechnology is presented by Ingham et al. [70]. They present the list of fields of AAO applications as follows:

- Counting and cell identification

- Growth and microcolony imagimg of microorganisms on AAO

- High-throughput microbiology

- Physical detection technologies

The unique properties of porous AAO are used in sensors and biosensors as already mentioned. A review article was published by Santos et al. [71] in which authors distinguished two types of AAO sensors: optical and electrochemical. The phenomena and techniques used in both types of sensors are summarized as follows:

Optical Biosensors

- Photoluminescence (PL) spectroscopy

- Surface-plasmon resonance (SPR)

- Waveguiding spectroscopy (WS)

- Localized surface-plasmon resonance (LSPR)

- Surface-enhanced Raman scattering spectroscopy (SERS)

- Reflectometric interference spectroscopy (RIfS)

Electrochemical Sensors and Biosensors

- Voltammetric and amperometric

- Impedance spectroscopy (IS)

- Capacitive, conductometric, and resistive sensors

Sensors and biosensors with AAO are used for the detection of gases, vapors, organic molecules, biomolecules, DNA, proteins, antibodies and cells, viruses, bacteria, cancer cells, which are in air, water, and biological environments.

As described in the previous chapter, complex composite systems produced by deposition inside the pores of AAO additional phases of metal, carbon, inorganic or organic compounds have been widely used in basic research and applications. This is especially due to their unique features like electrical, optical, catalytic, biological, and other properties. Major applications and references to the literature are given in Table 6.

No.	Application	Foreign phases in AAO matrix	References
1.	Membranes for ultra- and nanofiltration		[69,73]
2.	Sensors and biosensors		[71,72]
3.	Decorative coatings on aluminum	Sn, Ni, Cu	[20]
4.	Formation of metal nanoparticles and nanowires	Fe, Ni, Co, Au, Ag, Cu, Bi, Pt, Pd, Sn, W, Zn, Sb, Se, alloys	[66,74,75]
5.	Formation of dots and nanodots	InAs, SiGe, Si, CdTe, GaAs, Fe, Ag	[67,68,76,77]
6.	Formation of carbon structures, carbon nanotubes, graphene	C	[78-84]
7	Formation of inorganic compounds nanoparticles	GaN, ZnO, CdS, AgI, PbS	[85,86]
8.	Formation of organic compounds nanoparticles	Polypyrrole, polyaniline, DNA, polythiophene, ethylene vinyl acetate copolymer	[87,88]
9.	Examination of optical systems and models	Ni, Co	[64,89,90]
10.	Examination of magnetic systems and models	Fe, Ni, Co, alloys	[91-93]
11.	Catalysts of chemical reactions	Ni, Au, Pd, C	[81,94,95]
12.	Electronic components, capacitors, semiconductors	C, CdS, ZnO, compounds of In, Ga	[80,96-98]
13.	Luminescence, electroluminescence, and photoluminescence devices	Cu_2O, ZnS, In_2O_3	[96,99-102]
14.	Magnetic systems, data storage	Fe, Ni, Co, alloys Pt	[103-105]
15.	Sliding joints	MoS_2, Sn	[106]
16.	Solar collectors	Ni, Sn	[20]
17.	Solar batteries	CdS	[107,108]
18.	Lithium-ion batteries	$LiMn_2O_4$, CdS, CoSb	[109]
19.	Antibacterial materials	Ag	[110]
20.	Photonic crystals		[111,112]
21.	Fuel cells	Solid acid	[113]

Table 6. Main applications of AAO layers in science, technology, and nanotechnology

6. Summary

The structure of AAO layers can be precisely controlled with operation parameters of electrochemical oxidation. Complicated and expensive apparatus is not required for their preparation. It is for this reason that these layers are so widely used in nanotechnology.

Using a variety of techniques, in particular electrochemical methods, various composite materials of alumina and the foreign phase can be prepared with new or improved physical or chemical properties. AAO template having a uniform porous structure allows obtaining sustainable systems of nanoparticles, nanowires, nanotubes, and nanodots with an unlimited number of materials. AAO has become in recent years one of the most important materials for nanotechnology development, both in research as well as in applications.

The significant number of publications in this field now exceeds the size of one hundred items per year. Further applications of the AAO as nanofiltration membranes are anticipated, including drug delivery, gas permeation, and hemodialysis. Flat membranes are already commercially available and now efforts have been made to prepare tubular membranes [69].

In biomedical applications, AAO has been confirmed as a good substrate for the study of biomolecules, cells, and bacteria and their detection by means of optical imaging methods (SERS, optical waveguides), and fluorescence-based detection methods [70].

In the field of sensors and biosensors, AAO applications allow to broaden choice and miniaturization of sensors, especially those with optical measurements. The design of implantable biosensors with the ability to monitor biological systems in vivo and real time is promising for the application of AAO immunosensors [71-72].

It seems that in order to further the development of manufacturing methods and applications of metal nanostructures, it is advisable to develop further research in the following areas:

1. Basic research on the process of electrochemical oxidation of aluminum and of forming a specific porous structure of oxide layers.

2. Improving the methods of producing the best quality AAO with regular distribution of pores and desired pores geometry.

3. Further study of the effects of various factors on the structure and chemical composition of the AAO, such as:

   - Multistage oxidation

   - Use programmable waveform pulse currents

   - The importance of the pretreatment of the surface of aluminum substrate (alloy type, its structure, impurities)

   - Effect of heat pretreatment of the substrate and the final thermal treatment of AAO layer

4. Basic research of electrodeposition of metals in the pores of oxide layers.

5. The use of nonaqueous solutions or mixed solvents for metal deposition. In this way, the range of materials (metals and alloys) for nanostructures creations can be extended.

6. Study of the effect of complex current characteristics, including pulse current on electro-deposition processes of metals in the AAO template.

7. Examination of the physical (electrical, optical, magnetic, etc.) and chemical properties (catalytic, antibacterial, etc.); research on novel systems of metal nanoparticles and their alloys in a matrix of alumina or separated from oxide matrix.

Author details

Piotr Tomassi* and Zofia Buczko

*Address all correspondence to: piotr.tomassi@gmail.com

Institute of Precision Mechanics, Warsaw, Poland

References

[1] Romanowski W. *Highly Dispersed Metals*. Warszawa: PWN - Polish Scientific Publishers; 1987, Chichester: Ellis Horwood Limited Publishers; 1987, pp. 61-87.

[2] Feldheim DL, Colby AF.Jr. *Metal Nanoparticles. Synthesis, Characterization and Applications*. New York: Marcel Dekker, Inc.; 2002, pp. 119-139.

[3] Lin Q, Sun Zh. Study on optical properties of aggregated ultra-small metal nanoparticles. *Optik* 2011;122 1031-1036.

[4] Noguez C. Optical properties of isolated and supported metal nanoparticles. *Optic Mater* 2005;27 1204-1211.

[5] Wu B, Yuan H, Kuang A, Chen H, Zhang S. Static electric and optical properties of two coupled noble metal nanoparticles. *Comput Mater Sci* 2012;51 430-436.

[6] Kim JH, Yoon ChS. Modification of magnetic properties of metal nanoparticles using nanotemplate approach. *Thin Solid Films* 2008;516 4845-4850.

[7] Keller F, Hunter MS, Robinson DL. Structural features of oxide coatings on aluminium. *J Electrochem Soc* 1953;100(9) 411-419.

[8] Murphy JF, Michelson CE. A theory for the formation of anodic oxide coatings on aluminium. Proceedings of the Conference on Anodizing Aluminium, 1961, Nottingham UK, p. 83-95.

[9]   Palibroda E, Lupsan A, Pruneanu S, Saros M. Aluminium porous oxide growth. On the electric conductivity of the barrier layer. *Thin Solid Films* 1995;256(1-2) 101-105.

[10] Palibroda E. Aluminium porous oxide growth. II On the rate determining step. *Electrochim Acta* 1995;40(8) 1051-1055.

[11] Brace AW. *The technology of anodizing aluminium*, Interall S.r.l. Publishers, Modena, 2000, Italy, pp. 1-12.

[12] Wielage B, Alisch G, Lampke Th, Nickel D. Anodizing – a key for surface treatment of aluminium. *Key Engin Mater* 2008;384, 263-281.

[13] Parkhutik VP, Shershulsky VI. Theoretical modelling of porous oxide growth on aluminium. *J Physics D: Physics* 1992;25 1258-1263.

[14] Kanakala R, Singaraju PV, Venkat R, Das B. Modeling of porous alumina template formation under constant current conditions. *J Electrochem Soc*, 2005;152(1) 31-35.

[15] Singaraju P, Venkat R, Kanakala R, Das B. Model for porous alumina formation. Constant voltage anodization. *Euro Physic J, Appl Physics* 2006;35 107-111.

[16] Houser JE, Hebert HR. Modeling the potential distribution in porous anodic alumina films during steady-state growth. *J Electrochem Soc* 2006; 153(12) B566-B573.

[17] Wu Z, Richter C, Menon L, A study of anodization process during pore formation in nanoporous alumina templates. *J Electrochem Soc* 2007;154(1), E8-E12.

[18] Cheng Ch, Ngan AHW. Modelling and simulation of self-ordering in anodic porous alumina. *Electrochim Acta* 2011;56 9998-10008.

[19] Tomassi P. Mathematical model of anodic oxidation of aluminium. In: *Surface Treatment of Aluminium and other Light Metals*. Saulgau: Eugen G. Leuze Verlag; 1993. p. 14-18.

[20] Wernick S, Pinner R, Sheasby PG. *The Surface Treatment and Finishing of Aluminum and its Alloys*. Teddington: Finishing Publications Ltd.; 1987, pp. 289-368.

[21] Li XF. Chen D. Chen Zh. Wu Y, Wang ML, Ma N, Wang H. Stability of nanopore formation in aluminum anodization in oxalic acid. *Transact Nonferrous Mater Soc China* 2012;22 105-109.

[22] Han XY, Shen WZ. Improved two-step anodization technique for ordered porous anodic aluminum membranes. *J Electroanalytic Chem* 2011;655 56-64.

[23] Araoyinbo AO, Fauzi MNA, Sreekantan S, Aziz A. A novel process to produce nanoporous aluminum oxide using titration technique. *J Non-Crystalline Solids* 2010;356 1057-1060.

[24] Ono S, Saito M, Asoh H. Self-ordering of anodic porous alumina formed in organic acid electrolytes. *Electrochim Acta* 2005;51 827-833.

[25] Jia Y, Zhou H, Luo P, Chen J, Kuang Y. Preparation and characteristics of well-aligned macroporous films on aluminum by high voltage anodization. *Surf Coat Technol* 2006;201 513-518.

[26] Qin X, Zhang J, Meng X, Deng Ch, Zhang L, Ding G, Zeng H, Xu X. Preparation and analysis of anodic aluminum oxide films with continuously tunable interpore distances. *Applied Surface Science* 2015;328 459-465.

[27] Mozalev A, Mozaleva I, Sakairi M, Takahashi H. Anodic film growth on Al layers and Ta-Al metal bilayers in citric acid electrolytes. *Electrochimica Acta* 2005;50 5065-5075.

[28] Schwirn K, Lee W, Hillebrand R, Steinhart M, Nielsch K, Gosele U. Self-ordered anodic aluminum oxide formed by H_2SO_4 hard anodization. *ACS Nano* 2008;2(2) 302-310.

[29] Zaraska L, Sulka GD, Jaskuła M. Porous anodic alumina membranes formed by anodization of AA1050 alloy as templates for fabrication of metallic nanowire arrays. *Surf Coat Technol* 2010;205 2432-2437.

[30] Kirchner A, MacKenzie KJD, Brown IWN, Kemmitt T, Bowden ME. Structural characterisation of heat-treated anodic alumina membranes prepared using a simplified fabrication process. *J Membrane Sci* 2007;287 264-270.

[31] Bocchetta P, Sunseri C, Chiavarotti G, Di Quarto F. Microporous alumina membranes electrochemically grown. *Electrochim Acta* 2003;48 3175-3183.

[32] Zhao X, Seo SK, Lee UJ, Lee KH. Controlled electrochemical dissolution of anodic aluminum oxide for preparation of open-through pore structures. *J Electrochem Soc* 2007;154(10) C553-C557.

[33] Ginder RS, Hersam MC, Lipson AL. Unique pore-formation geometries in anodized aluminum oxide. *Nanoscope* 2010;7(1) 48-51.

[34] Nguyen TN, Kim D, Jeong DY, Kim MW, Kim JU. Formation behavior of nanoporous anodic aluminum oxide film in hot glycerol/phosphate electrolyte. *Electrochim Acta* 2012;83 288-293.

[35] Michalska-Domańska M, Norek M, Stępniowski WJ, Budner B. Fabrication of high quality anodic aluminum oxide (AAO) on low purity aluminum – a comparative study with the AAO produced on high purity aluminum. *Electrochim Acta* 2013;105 424-432.

[36] Li Y, Zheng M, Ma L, Shen W. Fabrication of highly ordered nanoporous alumina films by stable high-field anodization. *Nanotechnology* 2006;17 5101-5105.

[37] Lee W, Schwirn K, Steinhart M, Pippel E, Scholz R, Gosele U. Structural engineering of nanoporous anodic aluminum oxide by pulse anodization of aluminum. *Nanotechnology* 2008;3 234-239.

[38] Chung CK, Liao MW, Lee CT. Effects of temperature and voltage mode on nanoporous anodic aluminum oxide films by one-step anodization. *Thin Solid Films* 2011;520 1554-1558.

[39] Chung CK, Chang WT, Liao MW, Chang HC, Lee CT. Fabrication of enhanced anodic alumina oxide performance at room temperature using hybrid pulse anodization with effective cooling. *Electrochim Acta* 2011;56 6489-6497.

[40] Pashchanka M, Schneider JJ. Umiform contraction of high-aspect-ratio nanochannels in hexagonally patterned anodic alumina films by pulsed voltage oxidation. *Electrochim Acta* 2013;34 263-265.

[41] Su SH, Li CH, Zhang FB, Yokoyama M. Characterization of anodic aluminium oxide pores fabricated on aluminium templates. *Superlattices Microstruct* 2008;44 514-519.

[42] Sulka GD, Stępniowski WJ. Structural features of self-organized nanopore arrays formed by anodization of aluminum in oxalic acid at relatively high temperatures. *Electrochim Acta* 2009;54 3683-3691.

[43] Cui J, Wu Y, Wang Y, Zheng H, Xu G, Zhang X. A facile and efficient approach for pore-opening detection of anodic aluminum oxide membranes. *Appl Surf Sci* 2012;258 5305-5311.

[44] Zaraska L, Stępniowski WJ, Ciepiela E, Sulka GD. The effect of anodizing temperature on structural features and hexagonal arrangements of nanopores in alumina synthesized by two-step anodizing in oxalic acid. *Thin Solid Films* 2013;534 155-161.

[45] Zaraska L, Stępniowski WJ, Jaskuła M, Sulka GD. Analysis of nanopore arrangement of porous alumina layers formed by anodizing in oxalic acid at relatively high temperatures. *Appl Surf Sci* 2014;305 650-657.

[46] Asoh H, Ono S, Hirose T, Nakao M, Masuda H. Growth of anodic porous alumina with square cells. *Electrochim Acta* 2003;48 3171-3174.

[47] Choi J, Wehrspohn RB, Gosele U. Mechanism of guided self-organization producing quasi-monodomain porous alumina. *Electrochim Acta* 2005;50 2591-2595.

[48] Randon J, Mardilovich PP, Govyadinov AN, Paterson R. Computer simulation of inorganic membrane morphology Part 3. Anodic alumina films and membranes. *J Colloid Interf Sci* 1995;169 335-341.

[49] Sheasby PG, Cooke WE. The electrolytic colouring of anodized aluminium. *Transact Inst Metal Finish* 1974;52(3) 103-106.

[50] Herrmann E. Elektrolytisches Farben von anodisiertem Aluminium. *Galvanotechnik* 1972;63(2) 110-121.

[51] Sheasby PG, Patrie J, Badia M, Cheetham G. The colouring of anodized aluminium by means of optical interference effects. *Transact Inst Metal Finish* 1980;58(2) 41-47.

[52] Skominas V, Lichusina S, Miecinskas P, Jagminas A. A voltammetric and chronopotentiometric study of anodized aluminium in metal salt solutions. *Transact Inst Metal Finish* 2001;79(6) 213-218.

[53] Zemanova M, Gal M, Chovanceva M. Effect of frequency on pulse electrolytic colouring process of anodised aluminum. *Transact Inst Metal Finish* 2009;87(2) 97-101.

[54] Bograchev DA, Volgin VM, Davydov AD. Simple model of mass transfer in template synthesis of metal ordered nanowire. *Electrochimica Acta* 2013;96 1-7.

[55] Tomassi P. Electrochemical reactions to form composite layers aluminium oxide – metal. *Acta Physica Polonica A* 2002;102(2) 215-220.

[56] Ebling D. *Elektronische und Ionische Prozesse in Submikronstrukturen und deren Praparation im System Al/Al$_2$O$_3$*. PhD Thesis. Heinrich Heine University Dusseldorf; 1991. pp. 52-102.

[57] Hassel AW, Lohrengel MM. Initial stages of cathodic breakdown of thin anodic aluminium oxide films. *Electrochim Acta* 1995;40(4) 433-437.

[58] Hassel AW, Lohrengel MM, Schulze JW. Ultradunne Aluminiumoxid-Schichten. Untersuchung elektronischer und ionischer Transportprozesse. *Matalloberflache* 1996;50(1) 19-22.

[59] Montenero-Moreno JM, Belenguer M, Sarret M, Muller CM. Production of alumina templates suitable for electrodeposition of nanostructures using stepped techniques. *Electrochim Acta* 2009;54 2529-2535.

[60] Cheng W, Steinhart M, Gosele U, Wehrspon RB. Tree-like alumina nanopores generated in non-steady-state anodization. *J Mater Chem* 2007;17 3493-3495.

[61] Jani AMM, Losic D, Voelcker NH. Nanoporous anodic aluminium oxide: Advances in surface engineering and emerging applications. *Progress Mater Sci* 2013;58 636-704.

[62] Li AP, Muller F, Birner A, Nielsch K, Gosele U. Fabrication and microstructuring of hexagonally ordered two-dimensional nanopore arrays in anodic alumina. *Adv Mater* 1999;11(6) 483-487.

[63] Lee HJ, Kim DM, Park S, Lee Y, Koh WG. Micropatterning of a nanoporous alumina membrane with poly(ethylene glycol) hydrogel to create cellular micropatterns on nanotopographic substrates. *Acta Biomaterialia* 2011;7 1281-1289.

[64] Durtschi JD, Erali M, Herrmann MG, Elgort MG, Voelkerding KV, Smith RE. Optically improved aluminum oxide membrane through electroless Ni modification. *J Membrane Sci* 2005;248 81-87.

[65] Li X, Dong K, Tang L, Wu Y, Yang P, Zhang P. The fabrication of Ag nanoflake arrays via self-assembly on the surface of an anodic aluminum oxide template. *Appl Surf Sci* 2010;256 2856-2858.

[66] Zaraska L, Sulka GD, Jaskuła M. Porous anodic alumina membranes formed by anodization of AA1050 alloy as templates for fabrication of metallic nanowire arrays. *Surf Coat Technol* 2010;205 2432-2437.

[67] Gao T, Fan JC, Meng GW, Chu ZQ, Zhang LD. Thin Au film with highly ordered arrays of hemispheral dots. *Thin Solid Films* 2001;401 102-105.

[68] Vazquez M, Pirota KR, Navas D, Asenjo A, Hernandez-Velez M, Prieto P, Sanz JM. Ordered magnetic nanohole and antidot arrays prepared through replication from anodic alumina templates. *J Magnet Magnetic Mater* 2008;320 1978-1983.

[69] Lee KP, Mattia D. Monolithic nanoporous alumina membranes for ultrafiltration applications: Characterization, selectivity-permeability analysis and fouling studies. *J Membrane Sci* 2013;435 52-61.

[70] Ingham CJ, ter Maat J, de Vos WM. Where bio meets nano: The many uses for nanoporous aluminum oxide in biotechnology. *Biotechnol Adv* 2012;30 1089-1099.

[71] Santos A, Kumeria T, Losic D. Nanoporous anodic aluminum oxide for chemical sensing and biosensors. *Trends Analytic Chem* 2013;44 25-38.

[72] Koh G, Agarwal S, Cheow P.S, Tob CS. Development of membrane-based electrochemical immunosensor. *Electrochim Acta* 2007;53 803-810.

[73] Tomassi P, Buczko Z. Ultrafiltration membranes obtained by anodic oxidation of aluminium. *Proceedings of the V International Symposium "Forum Chemiczne '99'"*, Warszawa, 19-21 April, 1999. p. 181.

[74] Evans P, Hendren WR, Atkinson R, Wurtz GA, Dickson W, Zayats AV, Pollard RJ. Growth and properties of gold and nickel nanorods in thin film alumina. *Nanotechnology* 2006;17 5746-5753.

[75] Feliciano J, Martinez-Ifiesta MM. Synthesis and characterization of Pd, Cu and Ag nanowires in anodic alumina membranes using solid state reduction. *Mater Lett* 2012;82 211-213.

[76] Liang J., Luo H., Beresford R., Xu J. A growth pathway for highly ordered quantum dot arrays. *Appl Physics Lett* 2004;85 5974-5976.

[77] Jung M, Lee HS, Park HL, Mho S. Fabrication of high density CdTe/GaAs nanodot arrays using nanoporous alumina masks. *Curr Appl Physics* 2006;6S1 e187-e191.

[78] Ciambelli P, Arurault L, Sarno M, Fontorbes S, Leone C, Datas L, Sannino D, Lenormand P, Le Blond Du Plony S. Controlled growth of CNT in mesoporous AAO through optimized conditions for membrane preparation and CVD. *Nanotechnology* 2011;22(265613) 1-12.

[79] Jang WY, Kulkarni NN, Shih CK., Yao Zh. Electrical characterization of individual carbon nanotubes grown in nanoporous anodic alumina templates. *Appl Physics Lett* 2004;84(7) 1177-1179.

[80] Ahn HJ, Sohn JI, Lim YS, Shim HS, Kim WB, Seong TY. Electrochemical capacitors fabricated with carbon nanotubes grown within the pores of anodized aluminum oxide. *Electrochem Commun.* 2006;8(4) 513-516.

[81] Sigurdson S, Sundaramurthy V, Dalai AK, Adjaye J. Effect of anodic alumina pore diameter variation on template-initiated synthesis of carbon nanotube catalyst supports. *J Mol Cataly A: Chemical* 2009;306(1-2) 23-32.

[82] Jeong B, Uhm S, Kim JH, Lee J. Pyrolitic carbon infiltrated nanoporous alumina reducing contact resistance of aluminium/carbon interface. *Electrochim Acta* 2013;89(1) 173-179.

[83] Tu JP, Zhu LP, Hou K, Guo SY. Synthesis and frictional properties of array film of amorphous carbon nanofibres on anodic aluminium oxide. *Carbon* 2003;41(6) 1257-1263.

[84] Tu JP, Zhu LP, Hou K, Guo SY. Synthesis and frictional properties of array film of amorphous carbon nanofibres on anodic aluminium oxide. *Carbon* 2003;41(6) 125-1263.

[85] Cheng GS, Chen SH, Zhu XG, Mao YQ, Zhang LD. Highly ordered nanostructures of single crystalline GaN nanowires in anodic alumina membranes. *Mater Sci Engin* A 2000;286 165-168.

[86] Gomez H, Riveros G, Ramirez D, Henriquez R, Schrebler R, Marotti R, Dalchiele E. Growth and characterization of ZnO nanowire arrays electrodeposited into anodic alumina templates in DMSO solution. *Appl Physics A* 2012;16 197-204.

[87] Al-Kaysi RO, Bardeen CJ. General method for the synthesis of crystalline organic nanorods using porous alumina templates. *Chem Commun* 2006; 1224-1226.

[88] Cheng FL, Zhang ML, Wang H. Fabrication of polypyrrole nanowire and nanotube arrays. *Sensors* 2005;5 245-249.

[89] Li L. AC anodization of aluminum, electrodeposition of nickel and optical property examination. *Solar Energy Mater Solar Cells* 2000;64 279-289.

[90] Tang HJ, Wu FQ, Zhang S. Optical properties of Co/Al$_2$O$_3$ nano-array composite structure. *Appl Physics A* 2006;85 29-32.

[91] Metzger RM, Konovalov VV, Sun M, Xu T, Zangari G, Xu B, Benakli M, Doyle W.D. Magnetic nanowires in hexagonally ordered pores of alumina. *IEEE Transact Magnetics* 2000;36(1) 30-35.

[92] Qin DH, Lu M, Li H.L. Magnetic force microscopy of magnetic domain structure in highly ordered Co nanowire arrays. *Chem Physics Lett* 2001;350 51-56.

[93] Hamrakulov B, Kim IS, Lee MG, Park BH. Electrodeposited Ni, Fe, Co and Cu single and multilayer nanowire arrays on anodic aluminum oxide template. *Transact Nonferrous Metals Soc China* 2009 19 83-87.

[94]   Yu Y, Kant K, Shapter JG, Addai-Mensah J, Losic D. Gold nanotube membranes have catalytic properties. *Microporous Mesoporous Mater* 2012;153 131-136.

[95]   Pellin MJ, Stair PC, Xiong G, Elam JW, Birrell J, Curtiss L, George SM, Han CY, Iton L, Kung H, Kung M, Wang HH. Nesoporous catalytic membranes. Synthetic control of pore size and wall composition. *Cataly Lett* 2005;102 127-130.

[96]   Lei Y, Chim WK. Highly ordered arrays of metal/semiconductor core-shell nanoparticles with tunable nanostructures and photoluminescence. *J Am Chem Soc* 2005;127 1487-1492.

[97]   Mardare AI, Kaltenbrunner M, Sariciftci NS, Bauer S, Hassel AW. Ultra-thin anodic alumina capacitor films for plastic electronics. *Physica Status Solidi* A 2012;209 813-818.

[98]   Wen S, Mho S, Yeo IH. Improved electrochemical capacitive characteristics of the carbon nanotubes grown on the alumina templates with high pore density. *J Power Sources* 2006;163 304-308.

[99]   Li GH, Zhang Y, Wu YC, Zhang LD. Photoluminescence of anodic alumina membranes: pore size dependence. *Appl Physics A* 2005;81 627-629.

[100] Chen JH, Huang CP, Chao CG, Chen TM. The investigation of photoluminescence centres in porous alumina membranes. *App Physics A* 2006;84 297-300.

[101] Ko E, Choi J, Okamoto K, Tak Y, Lee J. Cu_2O nanowires in an alumina template: Electrochemical conditions for the synthesis and photoluminescence characteristics. *Chem Phys Chem* 2006;7 1505-1509.

[102] Li GH, Zhang Y, Wu YC, Zhang LD. Photoluminescence of anodic alumina membranes: pore size dependence. *Appl Physics A* 2005;81 627-629.

[103] Chu Sz, Inoue S, Wada K, Kurashima K. Fabrication of integrated arrays of ultrahigh density magnetic nanowires on glass by anodization and electrodeposition. *Electrochim Acta* 2005;51 820-826.

[104] Yasui K, Morikawa T, Nishio K, Masuda H. Patterned magnetic recording media using anodic porous alumina with single domain hole confoguration of 63 nm hole interval. *Jap J Appl Physics* 2005;44 L469-L471.

[105] Seo BI, Shaislamov UA, Lee SJ, Kim SW, Kim IS, Hong SH, Yang B. Growth of ferroelectric BLT and Pt nanotubes for semiconductor memories. *J Crystal Growth* 2006;292 315-319.

[106] Yu D, Feng Y, Zhu Y, Zhang X, Li B, Liu H.Template synthesis and characterization of molybdenum disulfide nanotubes. *Mater Res Bull* 2011;46 1504-1509.

[107] Aguilera A, Jayaraman V, Sanagapalli S, Suresh Singh R, Jayaraman V, Sampson K., Singh V.P. Porous alumina templates and nanostructured CdS for thin film solar cell applications. *Solar Energy Mater Solar Cells* 2006;90 713–726.

[108] Kang Y, Kim D. Well-aligned CdS nanorod/conjugated polimer solar cells. *Solar Energy Mater Solar Cells* 2006;90 166–174.

[109] Nishizawa M, Mukai K, Kuwabata S, Martin CR, Yoneyama H, Nishizawa M, Mukai K, Kuwabata S, Martin CR, Yoneyama H. Template synthesis of polypyrrole-coated spinel $LiMn_2O_4$ nanotubules and their properties as cathode active materials for lithum batteries. *J Electrochem Soc* 1997;144 1923-1927.

[110] Chi GJ, Yao SW, Fan J, Zhang WG, Wang HZ. Antibacterial activity of anodized aluminium with deposited silver. *Surf Coat Technol* 2002;157 162–165.

[111] Choi J, Luo Y, Wehrspohn RB, Hillebrand R, Schilling J, Gosele U. Perfect two-dimensional porous alumina photonic crystals with duplex oxide layers. *J Appl Physics* 2003;94 4757-4762.

[112] Masuda H, Ohya M, Nishio H, Asoh H, Nakao M, Nohtomi M, Tamamura T. Photonic crystal using anodic porous alumina. *Jap J Appl Physics* 1999;38 L1403-L1405.

[113] Bocchetta P, Chiavarotti G, Masi R, Sunseri C, Di Quarto F. Nanoporous alumina membranes filled with solid acid for thin film fuel cells at intermediate temperatures. *Electrochem Commun* 2004;6 923-928.

[114] Librant K. Formation of porous alumina by anodic oxidation process in sulfosalicylic acid solution. MSc Thesis. Technical University of Warsaw; 2011, pp. 38.

4

Electrodeposition of WO$_3$ Nanoparticles for Sensing Applications

L. Santos, J. P. Neto, A. Crespo, P. Baião, P. Barquinha, L. Pereira, R. Martins and E. Fortunato

Abstract

The motivation of using metal oxides is mainly due to its charge storage capabilities, and electrocatalytic, electrochromic and photoelectrochemical properties. But comparing with bulk, nanostructured materials present several advantages related with the spatial confinement, large fraction of surface atoms, high surface energy, strong surface adsorption and increased surface to volume ratio, which greatly improves the performances of these materials. The deposition of this materials can be accomplished by a variety of physical and chemical techniques but nowadays, electrodeposited metal oxides are generally used in both laboratories and industries due to the flexibility to control structure and morphology of the oxide electrodes combined with a reduced cost. Tungsten oxide (WO$_3$) is a well-studied semiconductor and is used for several applications as chromogenic material, sensor and catalyst. The major important features is its low cost and availability, improved stability, easy morphologic and structural control of the nanostructures, reversible change of conductivity, high sensitivity, selectivity and biocompatibility. For the electrodeposition of WO$_3$, more than one method can be adopted: electrodeposition from a precursor solution, anodic oxidation, and electrodeposition of already produced nanoparticles; however, in this case the mechanism of the electrodeposition is not fully understood. In this chapter, a review of the latest published work of electrodeposited nanostructured metal oxides is provided to the reader, with a more detailed explanation of WO$_3$ material applied in sensing devices.

Keywords: tungsten oxide, pH sensor, neural recordings, impedance

1. Introduction

Over the past two decades, the revolution in materials science has driven great advances in all areas of science and engineering. Nanoscience and nanotechnology are leading this revolution fueled by the industrial progress, the scientific ability to fabricate, model, and manipulate objects with a small number of atoms, and the continuous discovery of new phenomena at the nanoscale [1, 2]. Nanomaterials present unique properties, which are not found in the respective bulk materials [3]. Surface and quantum effects arise in nanostructures due to the large surface-to-volume ratio and to the dimensions that are comparable to the electron wavelength, respectively [4, 5].

In the metal oxides field, the discovery of superconductivity [6] and large magnetoresistance [7] has raised researchers' attention, especially to those with transition metals. Moreover, in traditional electronics, oxides are widely used as semiconductors, dielectrics, and conductive electrodes [8]. In the last years, nanostructured metal oxides for sensing applications have achieved significant advances, mainly due to their better thermal and environmental stability compared with organic materials. These devices, based on nanomaterials, can operate with low power consumption and can be easily integrated with nanoelectronics. Furthermore, the construction of sensors in "low-cost" substrates, such as plastic, paper, or textile, is also in demand for application in portable consumer devices [9–13]. Electrodeposition, in this case, is of great interest due to its flexibility to control the structure and morphology of the oxide electrodes combined with the reduced cost [14, 15].

2. Electrodeposition

The term electrodeposition is often used unclearly, referring either to electroplating or to electrophoretic deposition (EPD) [16]. The electroplating process is based on a solution of ionic species, usually in water, while EPD occurs in a suspension of particles. In electroplating, there is a charge transfer during the deposition to produce the metal or oxide layer in the electrode, while in EPD the deposition occurs without any reaction involved (Fig. 1). In fact, the principal driving force for EPD is the charge and the electrophoretic mobility of the particles in the solvent under the influence of an applied electric field, with the drawback that the solvent should be organic in order to avoid water electrolysis [16, 17].

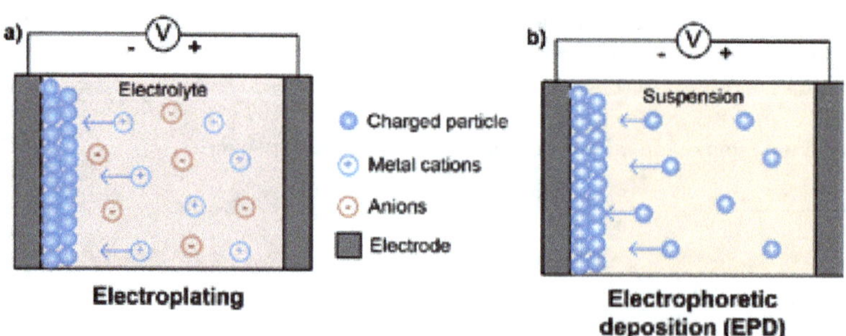

Figure 1. Schematic representation of the two types of cathodic electrodeposition processes: (a) electroplating and (b) electrophoretic deposition (EPD).

Another variation of an electrochemical deposition is the electroless (autocatalytic) deposition in which a reducing agent, dissolved in the electrolyte, is the electron source for the redox reaction, and no external power supply is needed [18]. Nevertheless, the electroless deposition will not be discussed in this chapter.

The first reports on the electrodeposition technique date back to the 19th century; however, the understanding of the process and the electrochemistry involved was only developed in the 20th century and it is believed that further research is still needed to optimize the process [16].

In electroplating, the relation between the current and the overpotential of electrodeposition is given by the Tafel equation (Equation 1), which describes the exponential dependence between the two parameters. Worth mentioning is that with the increase of the overpotential, the ionic current that the electrolyte can supply is limited either by material transport or electrical conductivity [15, 19].

$$i = -FkC \exp\left(\frac{\alpha F \eta}{RT}\right) \tag{1}$$

where i is the current, F is Faraday's constant, k a constant, C the concentration of metal ions in solution (which can be initially dissolved in the electrolyte or originated from the dissolution of the metallic anode), α the coefficient of symmetry ($\sim$ 0.5), η the overpotential, R the ideal gas constant, and T the absolute temperature (K).

The first attempt to correlate the amount of particles deposited by EPD with the different parameters influencing electrophoresis was first described by Hamaker for electrophoretic cells with a planar geometry. Over the years, Hamaker's law has been adapted and more recently Equation 2 was derived, relating the weight (W) of the charged particles deposited per unit area of electrode in the initial period, with different parameters, and disregarding the charge of the free ions [20].

$$W = \frac{2}{3}C\varepsilon_0\varepsilon_r\xi\mu^{-1}EL^{-1}t \tag{2}$$

Here, C is the concentration of the particles; ε_0 and ε_r the permittivity of vacuum and solvent, respectively; ξ the zeta potential of the particles; μ the viscosity of the solvent; E the applied potential; L the distance between the electrodes; and t the deposition time. Equation 2 demonstrates that the deposition weight of the charged particles under ideal EPD depends on all the previous parameters. However, if the solvent, the particles, and the apparatus for EPD are not changed, the weight of the deposited particles (W) is a function of C, E, and t. Therefore, the mass of the deposited particles, namely the thickness of the films, can be easily controlled by the concentration of the suspension, applied potential, and deposition time [17].

Electrodeposition of conventional metals for coatings has a very long history, with more than 200 years for some metals and alloys. Today, electrodeposition is much more than just a technique for coatings fabrication. In addition to applications such as decorative, wear, and corrosion-protective coatings, electrodeposition is also used for the manufacture of molds, functional coatings for magnetic and electronic applications, and microelectromechanical system components production [5]. In the future, though traditional applications will continue, new ones will rapidly develop, especially in the fields of nanoelectronics, biotechnology, and energy engineering. The electrodeposition of non-metallic materials will become more important and the combination of electrodeposition with other processes will lead to nano-structured materials with new and improved properties [21, 22]. Electrodeposition is extreme-ly versatile and different applications will keep being explored [23].

3. Metal oxide electrodeposition

Metal oxides are an important class of materials, which benefit from the large electronegativity of oxygen to induce strong bonding with nearby atoms [22]. At the same time, when compared with bulk materials, nanostructured metal oxides benefit from the spatial confinement, the large fraction of surface atoms, high surface energy, strong surface adsorption, and increased surface-to-volume ratio that greatly improves the performance of these materials [24].

The deposition of nanostructured metal oxides has been already reported by both physical and chemical methods [8, 5]. The advantages of electrodeposition include its speed, low cost, high purity, industrial applicability, use of different types of substrates, and production of films with different morphologies and compositions, as multilayers and alloys [21, 22].

In the electroplating of metal oxides, the reaction involved is usually defined by two consec-utive steps (Equation 3). First, the hydroxide will precipitate in the surface of the electrode due to the reaction of the metal ion (M^{n+}) in an alkaline solution, and secondly, the oxide is formed through a condensation/dehydration process. This last step can occur either during electro-deposition or by a subsequent annealing procedure [15].

$$M^{n+}_{(aq)} + nOH^{-}_{(aq)} \rightarrow M(OH)_{n(ads)} \rightarrow MO + nH_2O \tag{3}$$

Another alternative is the formation of metal oxides by anodic oxidation [15]. In this case, the source of the metal ions is the metallic anode and the metal oxide film will be deposited on top of the metal electrode. The general equation can be described as (Equation 4):

$$M + mH_2O \leftrightarrow MO_{m(s)} + 2mH^{+}_{(aq)} + 2me^{-} \tag{4}$$

In EPD, the metal oxide nanoparticles are generally synthesized by different solution based techniques, e.g., sol-gel, precipitation, and hydrothermal synthesis, prior to deposition. The main challenge of this technique is the preparation of a stable dispersion that originates a film with good properties, uniformity, and appropriate thickness. The use of dispersants, binders, or other additives that influences the agglomeration and charge of the particles contributes to the tuning of the properties of the deposited film and need to be considered in defining the EPD parameters [16].

3.1. Applications

Nowadays, electrodeposited nanostructured metal oxides are generally used for different applications in laboratories and industry [15]. The latest published reports on the field, listed in Table 1, evidentiate the diversity of areas where these materials can be applied, as presented below.

The deposition of metal/metal oxide nanoparticles composites allowed advances on the protective coatings field. Sajjadnejad et al. [25] improved the corrosion resistance of zinc by co-depositing TiO_2 nanoparticles, while Zeng et al. [26] incorporated CeO_2 nanoparticles to improve the corrosion behavior of nickel coatings. Charlot et al. [27] opened the discussion of the kinetics and mechanism of the anodic EPD of SiO_2 nanoparticles to improve the control of the thickness and properties of these coatings (Fig. 2).

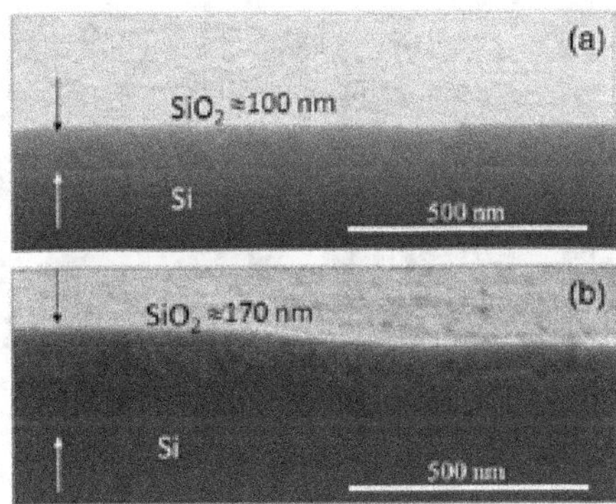

Figure 2. Scanning electron microscopy (SEM) images of a film cross-section obtained from a suspension with a mass fraction of 3% of nanoparticles under an applied electric field of (a) 6 V cm^{-1} and of (b) 60 V cm^{-1}. Reprinted from [27], with permission from Elsevier.

Metal oxide nanostructures are already known to show good catalytic properties. Tu et al. [28] produced Cu_2O-Cu nanoparticles in carbon paper via electroplating. This procedure is an easy, one-step technique that can be an attractive candidate as a visible-light-driven photocatalyst. At the same time, Yoon et al. [29] studied the influence of 2D and 3D structures on electrodeposited Cu_2O films by controlling the electrolyte pH and by using polystyrene (PS) beads as template, respectively. This techniques allowed the production of electrodes with increased surface area. Battaglia et al. [30] also improved the catalytic performance of different Ni electrodes by electrodepositing IrO_2 nanostructures through different electrochemical methods. The composites obtained by galvanostatic deposition of the oxide catalyst presented the best activity for water splitting applications.

Solid oxide fuel cells (SOFC) have shown to be a good alternative for electric power generation systems. SOFC show high energy conversion efficiency, clean power generation, reliability, modularity, fuel adaptability, noise-free, excellent long-term stability, and versatility for direct conversion of chemical energy to electrical energy. In this field, Das and Basu [31] applied the EPD technique to deposit yttria-stabilized zirconia (YSZ) nanoparticles on a NiO-YSZ substrate, which after sintering was suitable for application in SOFC (Fig. 3).

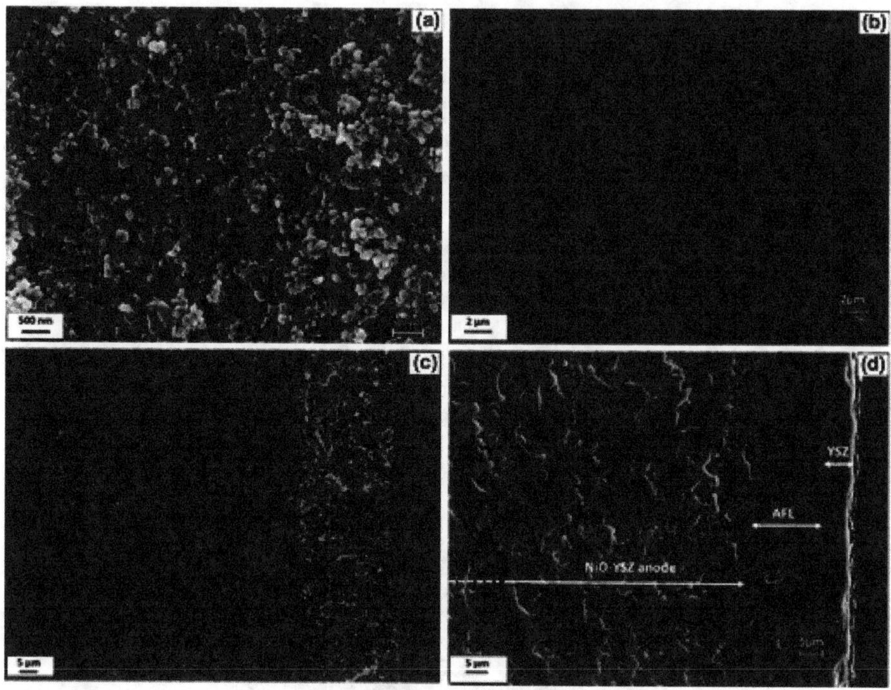

Figure 3. Field-emission SEM images of top view and cross-section of yttria-stabilized zirconia (YSZ) electrophoretic deposition coating (a) (c) as-deposited and (b) (d) sintered at 1400 °C for 6 h, directly deposited onto the conducting polymers such as polypyrrole-coated NiO-YSZ substrate at a constant applied voltage of 15 V. Reprinted from [31], with permission from John Wiley and Sons.

EPD was also the technique used to deposit TiO_2 nanoparticles for dye-sensitized solar cells (DSSC) [32] and Li-ion micro-batteries applications [33]. For DSSC, the thickness of the TiO_2 films was controlled by changing the deposition time and the I_2 dosage that electrically charge the nanoparticles, while for batteries, the EPD was performed with different TiO_2 structures and different 3D aluminum collectors configurations (Fig. 4). The effect of the substrate was also tested in the EPD of ZnO nanoparticles for conductive fabrics applications [34]. Liu et al. [35] studied the EDP of metal oxides using celestine blue as charging and dispersing agent. The nanostructured MnO_2 films were applied for energy storage in electrochemical supercapacitors with high capacitance and excellent capacitance retention at high charge-discharge rates.

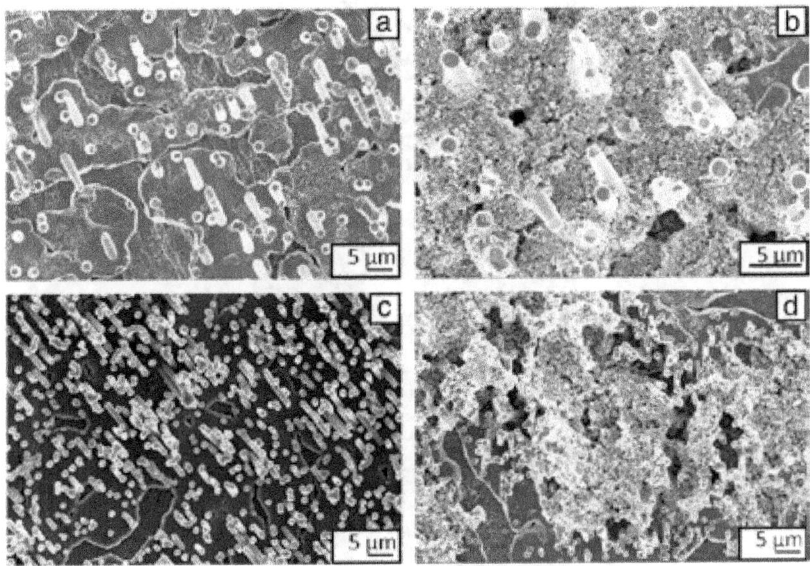

Figure 4. SEM micrographs of aluminium rods obtained by pulsed galvanostatic deposition using a PC membrane with (a) 2 µm-pore size and (c) 1 µm-pore size with electrophoretically deposited P25 particles attached to the respective 3D Al substrates (b) and (d). Reprinted from [33], with the permission from Elsevier.

The use of metal oxides offers functionalities that vary from electrically conducting to insulating and from highly catalytic to inert, which are useful for sensing applications. Different types of metal oxide sensors have been investigated for several decades, and it has been proved that the reduction of crystallite size provided a significant increase in the sensing performances. Even if less established, these type of sensors are very promising and new developments are being accomplished every day [36].

Recently, Cu_2O nanostructures were electroplated to produce a facile and economic photoelectrochemical sensor [37], while Ir_2O_3 was deposited in stretchable and multiplexed pH sensors [38]. This sensor combined electrochemical, microfabrication, and printing techniques and was successfully applied in beating explanted cardiac tissue, with accurate spatiotemporal monitoring of changes in pH (Fig. 5).

Monitoring analgesic drugs with the use of biosensors allows a rapid, reliable, and sensitive method without the requirement of a sample pre-treatment. For that, alloys deposition allows the combination of different materials properties without compromising thickness or surface area available. The biosensors developed by Narang et al. [39] were produced by EPD of an Fe_2O_3 magnetic nanoparticle coated with ZrO suspension containing chitosan, prior to enzyme (horseradish peroxidase) immobilization. Also the combination of Fe_2O_3 with carbon nanotubes and chitosan was earlier used by Batra et al. [40] to immobilize hemoglobin and were applied as an amperometric biosensor.

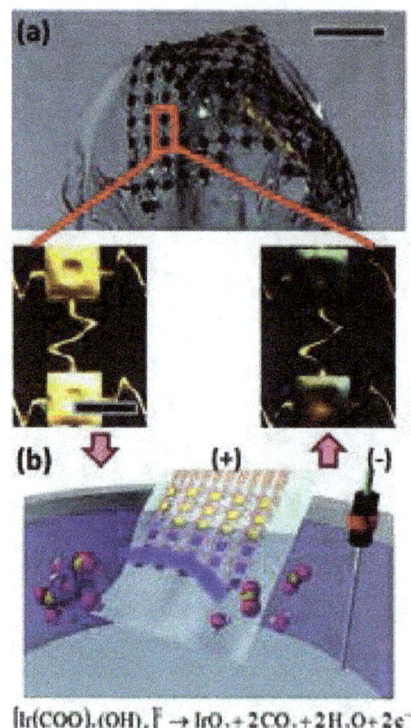

Figure 5. (a) Picture of the produced pH sensors with the magnified images of the gold electrodes before (lower left) and after (lower right) IrO$_x$ electroplating. The scale bars correspond to 5 and 0.5 mm for the upper and the lower images, respectively. (b) Schematic illustration of the chemical reactions during IrO$_x$ electroplating. Reprinted from [38], with permission from John Wiley and Sons.

Application	Nanomaterials/Composites	References
Corrosion and wear resistive coatings	Zn-TiO$_2$, Ni-CeO$_2$, SiO$_2$	[25][26][27]
Photocatalyst	Cu$_2$O-Cu, Cu$_2$O	[28][29]
Water splitting	Ni-IrO$_2$	[30]
Solid oxide fuel cell	Y$_2$O$_3$-ZrO$_2$ (YSZ)	[31]
Dye-sensitized solar cell	TiO$_2$	[32]
Li-ion micro-battery	TiO$_2$	[33]
Conductive fabric	ZnO	[34]
Supercapacitor	MnO$_2$	[35]
Photoelectrochemical sensor	Cu$_2$O	[37]
pH sensor	Ir$_2$O$_3$	[38]
Biosensor	ZrO@Fe$_3$O$_4$, cMWCNT-Fe$_3$O$_4$	[39][40]

Table 1. List of the latest published research on electrodeposited metal oxide nanostructures/nanomaterials.

4. Nanostructured WO_3

Tungsten oxide (WO_3) is a well-studied semiconductor used for several applications such as chromogenic material, sensor, and catalyst [41]. The major advantages is its low cost and availability, improved stability, reversible change of conductivity and optical properties, high sensitivity, selectivity, and biocompatibility [42].

Transition-metal oxides, especially those with d^0 and d^{10} electronic configurations, as WO_3, TiO_2, or ZnO show interesting properties and stability that are important for sensing applications [43]. The energy band gap of WO_3 corresponds to the difference between the energy levels of the valence band formed by the filled O 2p orbitals and the conduction band formed by empty W 5d orbitals, ranging from 2.6 to 3.25 eV [44]. In nanostructured WO_3, the bandgap generally increases with the reduction of the grain size, which is attributed to the quantum confinement effect [45]. Tungsten oxide is also well known for its properties in a non-stoichiometric form, since its lattice can support a significant concentration of oxygen vacancies [44].

4.1. WO_3 electrodeposition

Many liquid and vapor phase synthesis methods have been used to synthesize WO_3 [45]. Nevertheless, for the electrodeposition of nanostructured WO_3 films, more than one method can be adopted: electroplating from a precursor solution [46, 47], anodic oxidation from a metal layer [47–49], and electrodeposition from a WO_3 nanoparticles dispersion [50, 51]. A list of the latest reports is presented in Table 2.

WO_3 Precursor	Nanostructured film	Application	References
Na_2WO_4	WO_3	--	[59]
Na_2WO_4	$Pt-WO_3$	Proton exchange membrane fuel cell	[53]
Na_2WO_4	TiO_2-WO_3	Photocatalyst	[60]
Na_2WO_4	TiO_2-WO_3	Water splitting	[61]
H_2WO_4	WO_3/PANI	Supercapacitor	[52]
PTA	WO_3	Electrochromic film	[46]
PTA	WO_3/PEDOT	Electrochromic film	[62]
W	WO_3	--	[57]
W	WO_3/PANI	Electrocatalyst	[48]
W	TiO_2-WO_3	Photoelectrocatalyst	[63]
W	WO_3	Photoelectrocatalyst	[64]
W	WO_3	Photocatalyst	[65]
W	NH_4-doped WO_3	Water splitting	[49]
W	TiO_2-WO_3	Water splitting	[66]

WO$_3$ Precursor	Nanostructured film	Application	References
W	WO$_3$ – TiO$_2$	Electrochromic film	[54]
W	WO$_3$	H$_2$ sensor	[67]
WO$_3$ NPs	WO$_3$/henna	Dye sensitized solar cells	[68]
WO3 NPs	WO$_3$	Water splitting	[51]
WO$_3$ NWs	WO$_3$	Electrochromic film	[58]
WO$_3$ NPs	WO$_3$	Electrochromic film	[50]
WO$_3$ NPs	WO$_3$	pH sensor	[12]
WO$_3$ NPs	WO$_3$	Neural electrodes	[69]

Table 2. Resume of the latest published research on electrodeposited nanostructured WO$_3$ with the respective precursors and final applications.

Cathodic electroplating is usually based on the local increase of the pH near the electrode surface due to the reduction of O$_2$ or H$_2$O, which induces precipitation of metal ions present in the solution as metal oxide or hydroxide. For the deposition of WO$_3$, the reactions involved in the formation of the oxide are usually based on the formation of the peroxytungstate (W$_2$O$_{11}^{2-}$) intermediate from a tungstate salt (or from the reaction of metallic tungsten with hydrogen peroxide), as described in Equations 5 and 6 [46, 52].

$$2WO_4^{2-} + 4H_2O_2 + 2H^+ \rightarrow W_2O_{11}^{2-} + 5H_2O \tag{5}$$

$$W_2O_{11}^{2-} + 2H^+ \rightarrow 2WO_3 + 2O_2 + H_2O \tag{6}$$

Depending on the electrochemical potential and solution pH, the WO$_3$ phase may also be involved in other reactions, as the formation of sub-stoichiometric oxide and tungsten bronze (Equations 7 and 8) or even re-dissolution of the oxide phase (Equation 9). The reduced phases formed by these reactions have higher conductivity and hydrophilicity than WO$_3$ and should be considered during characterization of the deposited films [53].

$$WO_3 + 2yH^+ + 2ye^- \leftrightarrow WO_{3-y} + yH_2O \tag{7}$$

$$WO_3 + xH^+ + xe^- \leftrightarrow H_xWO_3 \tag{8}$$

$$WO_3 + H_2O \rightarrow WO_4^{2-} + 2H^+ \tag{9}$$

For the anodic oxidation procedure, the general equation can be expressed as Equation 10 [54] and the full mechanism is explained by the occurrence of different reactions simultaneously, as the synthesis of surface oxide films (e.g., W_2O_5, WO_2) and tungstate ions (WO_4^{2-}) [55, 56]. This oxidation is usually followed by the slow dissolution of the oxide phase, as in Equations 11 or 12 depending on the solution pH [55, 57].

$$W + 3H_2O \leftrightarrow WO_3 + 6H^+ + 6e^- \tag{10}$$

$$WO_3 + 2H^+ \rightarrow WO_2^{2+} + H_2O \tag{11}$$

$$WO_3 + 2OH^- \rightarrow WO_4^{2-} + H_2O \tag{12}$$

In the case of the deposition from WO_3 nanoparticles dispersions (EPD), the mechanism is not yet fully understood. The majority of the authors agree that the deposition occurs through an electrophoretic mechanism driven by the surface charge of the particles [51, 58], but in fact, the potential (or current) applied during deposition can also promote tungsten reduction from W^{6+} to W^{5+} that is counterbalanced by the cation intercalation into the oxide structure, as described in Equation 8, thus forming tungsten bronze (H_xWO_3) [50]. In the work of Liu et al. [50], XRD and optical characterization showed that HWO_3 was obtained as the main phase of the deposited films, which supports the hypothesis of the mechanism via electrochemical reduction. Furthermore, since the reduced WO_3 is significantly more conductive than the oxidized form, it allowed continuous film growth. In the future, further analysis of the deposited films should be conducted to confirm the electrochemical deposition mechanism.

4.2. WO$_3$ sensing applications

4.2.1. Gas sensors

Precise and affordable monitoring of chemical gases is a critical issue for human health, industrial processes, and environmental protection. For that, nanostructured WO_3 has been intensively studied due to its excellent sensing capabilities and reproducibility. These characteristics are mainly ascribed to the increased surface area and complete depletion of carriers within the nanostructure when exposed to the target gas [45]. The gas sensing mechanism is described by the increase or decrease of the conductance of the oxide layer when exposed to reducing (H_2, H_2S, CO) or oxidizing (NO_2, O_3, CO_2) gases, respectively.

$$H_2 + O_{ads}^- \rightarrow H_2O + e^- \tag{13}$$

In Equation 13, H_2 adsorbs and reacts with O^- formed on the surface of the electrode, increasing the surface conductance and releasing the captured electrons [67].

$$NO_2 + e^- \rightarrow NO_2^- \tag{14}$$

$$2NO_2 + 2O_{ads}^- \rightarrow 2NO_2^- + O_2 \tag{15}$$

When NO_2 is targeted on the WO_3 surface, it not only reacts with the electrons from the conduction band (Equation 14) but also with the chemisorbed oxygen (Equation 15), thus promoting a depletion on the surface of the electrode and, consequently, the increase on resistance [70, 71].

An example of a hydrogen gas sensor was built by Yang et al. [67] through anodic oxidation of a tungsten layer previously deposited by radio frequency magnetron sputtering on a sapphire substrate (Fig. 6). The nanoporous WO_3 film sensor, after annealing at 600°C, exhibited good sensitivity to H_2 gas in air.

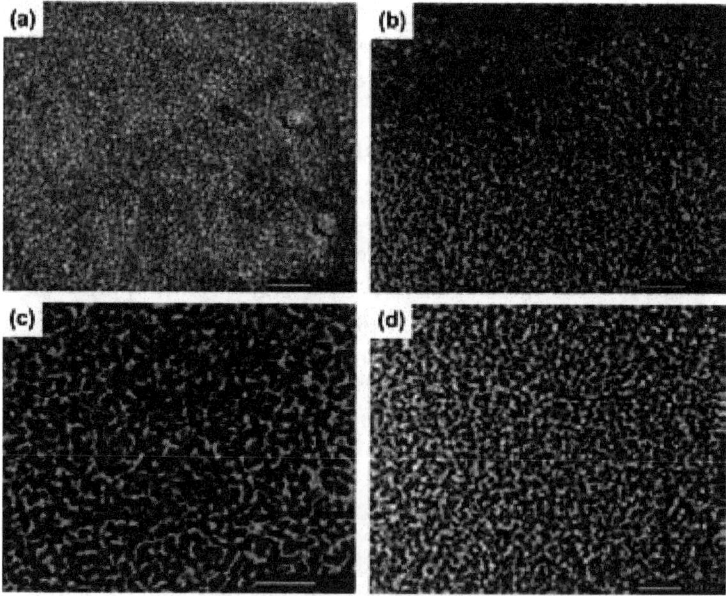

Figure 6. SEM images of tungsten oxide films with different anodic oxidation voltages: (a) 20 V, (b) 30 V, (c) 50 V, and (d) 60 V operating at an electrode distance of 2 cm for 60 min. Reprint from [67], with permission from Cambridge University Press.

4.2.2. Biosensors

The application of WO_3 to other sensing platforms, as in biosensors, is mainly due to the electrical and optical properties mentioned above [72]. In fact, it was already demonstrated that nanoparticles of metal oxides applied to suitable electrode surfaces allow protein immobilization and biocatalytic processes to be driven electrochemically [73]. However, to the best of the authors' knowledge, only Feng et al. [74] employed electrodeposited nanostructured

WO$_3$ films to enhance the hemoglobin protein loadings, accelerate interfacial electron transfer, and improve thermal stability of the adsorbed protein. The influence of the electrodeposition time to the response time and peak current of the electrode is demonstrated in Fig. 7.

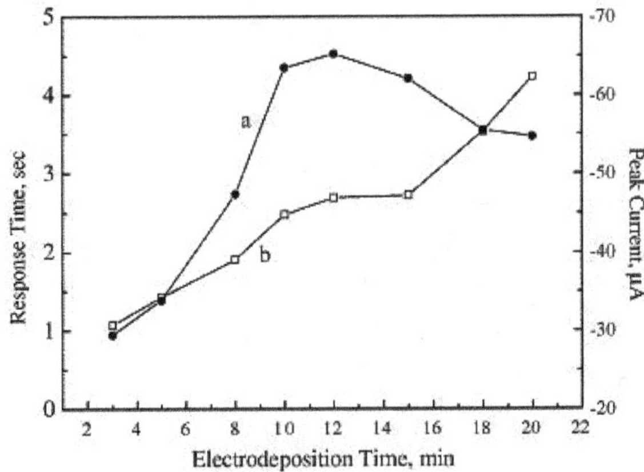

Figure 7. Influence of electrodeposition time on (a) peak current of the cyclic voltammograms in phosphate buffer solution (PBS, pH 6.0) at 100 mV s^{-1} and (b) typical steady-state response time of Hb/meso-WO$_3$/graphite electrodes. Reprinted from [74], with permission from Elsevier.

4.2.3. pH sensors

The pH value can be used as an indicator for disease diagnostics, medical treatment optimization, and monitoring of biochemical and biological processes [75]. Nevertheless, the integration of pH sensing systems into the next generation of wearable devices requires a different architecture than currently used in typical glass-type electrodes and a minimal electrode size [76]. In addition, technological and industrial efforts are under way to incorporate different sensors into our daily life by assembling these sensors on common substrates such as plastic, textile, and paper [9]. In the work reported earlier [12], flexible pH sensors were based on electrodeposited WO$_3$ sensing layer in a gold/polyimide substrate (Fig. 8). The pH sensing mechanism for this material, even if not fully understood, is believed to be dependent of the redox reaction involving the production of the tungsten bronze with a higher conductivity than the tungsten oxide (Equation 8).

4.2.4. Neural electrodes

Microtechnology allowed the arrangement of multiple microelectrodes on the same substrate over small distances (Fig. 9a). Nevertheless, in order to provide sufficient recording sensitivity to small electrodes for measuring neuron electrical activity, they are often coated with different nanostructured or conducting materials to increase the effective surface area and electrochemical interface capacitance [77–79]. The interest in utilizing transition metal oxide films is due to its pseudocapacitive character related to chemisorption processes and redox reactions that take place at the surface [80]. Since nanostructured WO$_3$ has already proved to enhance capacitive performances due to its large surface area and low charge transport resistance [52],

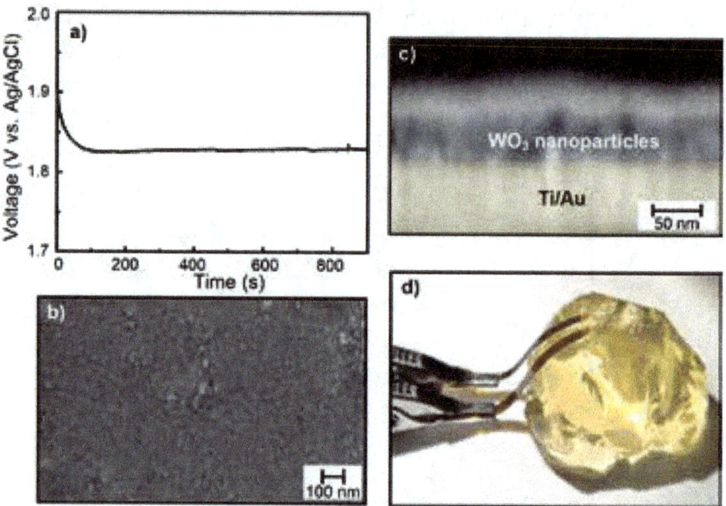

Figure 8. (a) Voltage response during electrodeposition at 20 µA; (b) topographic and (c) cross-section SEM images of the WO$_3$ electrodeposited layer; and (d) photograph of the prototype WO$_3$ sensor using a flexible Ag/AgCl reference electrode in a non-planar surface made of gelatin-based electrolyte [12].

it was used for neural recordings applications [69]. The optimization of the electrodeposition parameters led to a slight increase on the charge storage capacity ($\sim$10%) and a decrease of the impedance values, of approximately 40% (Fig. 9b and 9c).

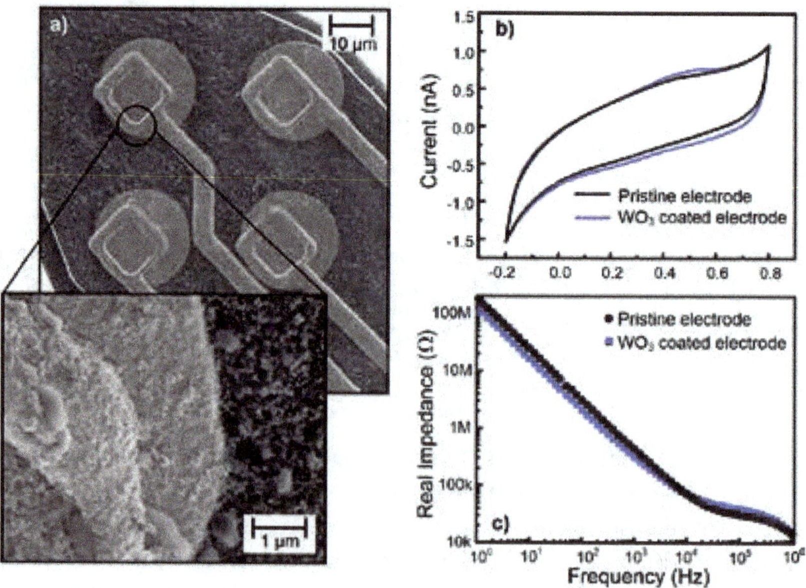

Figure 9. (a) SEM images of the Neuronexus electrode and a detail of the iridium electrode (lighter area) coated with WO$_3$ nanoparticles, electrodeposited at 30 nA for 15 s; (b) cyclic voltammetry and (c) electrochemical impedance characterizations of the pristine (black) and coated electrodes (blue) [69].

These preliminary results show the versatility of electrodeposition in different materials and configurations as well as in different sensing mechanisms.

5. Conclusions

Tungsten oxide (WO_3) is one of the most studied metal oxide and the sensing performance of this material is of great interest due to the capability of reversible change of both its optical and electrical properties. The evolution in the fields of nanoscience and nanotechnology allowed these materials to replace many organic and metallic materials in a huge range of applications besides creating new areas of development. The increased surface area and the quantum confinement effects in size ranges below 100 nm make nanostructured WO_3 a good platform for gas and pH sensors, along with neural electrodes and biosensors.

In the last decade, the use of electrodeposition for nanostructured metal oxide films has been growing due to the versatility of this method in different applications and materials. Just in the last year, applications varied from catalysts and sensors to capacitors. The use of different types of templates and the deposition of composites will contribute to the continued development of this technique.

Acknowledgements

The authors would like to thank the Portuguese Science Foundation (FCT-MEC) through project EXCL/CTM-NAN/0201/2012, Strategic Project UID/CTM/500025/2013, and doctoral grants SFRH/BD/73810/2010 given to L. Santos and SFRH/BD/76004/2011 given to J. Neto. The authors would like to thank Dr. Adam Kampff from Champalimaud Center of Unknown for the knowledge transfer related with neural electrodes.

Author details

L. Santos[1*], J. P. Neto[1,2], A. Crespo[1], P. Baião[1,2], P. Barquinha[1], L. Pereira[1], R. Martins[1] and E. Fortunato[1]

*Address all correspondence to: ls.santos@campus.fct.unl.pt; emf@fct.unl.pt

1 CENIMAT/I3N, Departamento de Ciências dos Materiais, Faculdade de Ciências e Tecnologia, Universidade Nova de Lisboa and CEMOP/Uninova, Caparica, Portugal

2 Champalimaud Centre for the Unknown, Champalimaud Neuroscience Programme, Lisbon, Portugal

References

[1] Koch CC, Ovid'ko IA, Seal S, Veprek S. Structural Nanocrystalline Materials: Fundamentals and Applications. Cambridge, UK: Cambridge University Press; 2007.

[2] Roco MC. Broader societal issues of nanotechnology. J Nanoparticle Res 2003;5:181–9.

[3] Schmid G, editor. Nanoparticles: From theory to application. Weinheim, Germany: WILEY-VCH Verlag GmbH & Co. KGaA; 2004.

[4] Society R. Nanoscience and Nanotechnologies: Opportunities and Uncertainties. London, UK: The Royal Society & The Royal Academy of Engineering; 2004.

[5] Koch CC, editor. Nanostructured Materials: Processing, Properties, and Applications. Norwich, NY, USA: William Andrew Publishing; 2007.

[6] Cava RJ. Oxide Superconductores. J Am Ceram Soc 2000;83:5–28.

[7] Sun JZ, Gupta A. Spin-Dependent Transport and Low-Field Magnetoresistance in Doped Manganites. Annu Rev Mater Sci 1998;28:45–78. doi:10.1146/annurev.matsci.28.1.45.

[8] Barquinha P, Martins R, Pereira L, Fortunato E. Transparent Oxide Electronics. Chichester, UK: John Wiley & Sons, Ltd; 2012. doi:10.1002/9781119966999.

[9] Hu B, Chen W, Zhou J. High performance flexible sensor based on inorganic nanomaterials. Sensors Actuators B Chem 2013;176:522–33. doi:10.1016/j.snb.2012.09.036.

[10] Costa MN, Veigas B, Jacob JM, Santos DS, Gomes J, Baptista P V, et al. A low cost, safe, disposable, rapid and self-sustainable paper-based platform for diagnostic testing: Lab-on-paper. Nanotechnology 2014;25:094006. doi:10.1088/0957-4484/25/9/094006.

[11] Branquinho R, Pinto J V, Busani T, Barquinha P, Pereira L, Baptista PV, et al. Plastic Compatible Sputtered Ta2O5 Sensitive Layer for Oxide Semiconductor TFT Sensors. J Disp Technol 2013;9:723–8.

[12] Santos L, Neto JP, Crespo A, Nunes D, Costa N, Fonseca IM, et al. WO3 Nanoparticle-Based Conformable pH Sensor. ACS Appl Mater Interfaces 2014;6:12226–34. doi:10.1021/am501724h.

[13] Pinto JV, Branquinho R, Barquinha P, Alves E, Martins R, Fortunato E. Extended-Gate ISFETs Based on Sputtered Amorphous Oxides. J Disp Technol 2013;9:729–34. doi:10.1109/JDT.2012.2227298.

[14] Rao CNR, Muller A, Cheetham AK, editors. The chemistry of nanomaterials: Synthesis, Properties and Applications. Vol. 1. Weinheim, Germany: WILEY-VCH Verlag GmbH & Co. KGaA; 2004.

[15] Garcia EM, Lins VFC, Matencio T. Metallic and Oxide Electrodeposition. In: Aliofkhaz-raei M, editor. Mod. Surf. Eng. Treat., InTech; 2013.

[16] Besra L, Liu M. A review on fundamentals and applications of electrophoretic deposition (EPD). Prog Mater Sci 2007;52:1–61. doi:10.1016/j.pmatsci.2006.07.001.

[17] Dickerson JH, Boccaccini AR, editors. Electrophoretic Deposition of Nanomaterials. Springer; 2012.

[18] Paunovic M, Schlesinger M. Fundamentals of Electrochemical Deposition. 2nd editio. Hoboken, NJ, USA: John Wiley & Sons, Inc.; 2006.

[19] Schlesinger M, Paunovic M, editors. Modern Electroplating. Hoboken, NJ, USA: John Wiley & Sons, Inc.; 2010.

[20] Chen F, Liu M. Preparation of yttria-stabilized zirconia (YSZ) films on La0.85Sr0.15MnO3 (LSM) and LSM-YSZ substrates using an electrophoretic deposition (EPD) process. J Electrochem Soc 2001;21:127–34.

[21] Gurrappa I, Binder L. Electrodeposition of nanostructured coatings and their character-ization—a review. Sci Technol Adv Mater 2008;9:043001. doi: 10.1088/1468-6996/9/4/043001.

[22] Djoki S, editor. Electrodeposition and Surface Finishing. New York, USA: Springer; 2014.

[23] Schwarzacher W. Electrodeposition: A Technology for the Future. Electrochem Soc Interface 2006;15:32–5.

[24] Edelstien AS, Cammarata RC, editors. Nanomaterials—Synthesis, Properties and Ap-plications. London, UK: IOP Publishing Ltd; 1996.

[25] Sajjadnejad M, Ghorbani M, Afshar A. Microstructure-corrosion resistance relation-ship of direct and pulse current electrodeposited Zn-TiO2 nanocomposite coatings. Ceram Int 2015;41:217–24. doi:10.1016/j.ceramint.2014.08.061.

[26] Zeng YB, Qu NS, Hu XY. Preparation and Characterization of Electrodeposited Ni-CeO2 Nanocomposite Coatings with High Current Density. Int J Electrochem Sci 2014;9:8145–54.

[27] Charlot A, Deschanels X, Toquer G. Submicron coating of SiO2 nanoparticles from electrophoretic deposition. Thin Solid Films 2014;553:148–52. doi:10.1016/j.tsf.2013.11.064.

[28] Tu J, Yuan Y, Jiao H, Jiao S. Controllable Cu2O–Cu nanoparticle electrodeposition onto carbon paper and its superior photoelectrochemical performance. RSC Adv 2014;4:16380. doi:10.1039/C4RA00592A.

[29] Yoon S, Kim M, Kim I-S, Lim J-H, Yoo B. Manipulation of cuprous oxide surfaces for improving their photocatalytic activity. J Mater Chem A 2014;2:11621. doi:10.1039/C4TA00616J.

[30] Battaglia M, Inguanta R, Piazza S, Sunseri C. Fabrication and characterization of nanostructured Ni–IrO2 electrodes for water electrolysis. Int J Hydrogen Energy 2014;39:16797–805. doi:10.1016/j.ijhydene.2014.08.065.

[31] Das D, Basu RN. Electrophoretic Deposition of Zirconia Thin Film on Nonconducting Substrate for Solid Oxide Fuel Cell Application. J Am Ceram Soc 2014;97:3452–7. doi:10.1111/jace.13163.

[32] Chou J-C, Lin S-C, Liao Y-H, Hu J-E, Chuang S-W, Huang C-H. The Influence of Electrophoretic Deposition for Fabricating Dye-Sensitized Solar Cell. J Nanomater 2014;2014:1–7. doi:10.1155/2014/126053.

[33] Oltean G, Valvo M, Nyholm L, Edström K. On the electrophoretic and sol–gel deposition of active materials on aluminium rod current collectors for three-dimensional Li-ion micro-batteries. Thin Solid Films 2014;562:63–9. doi:10.1016/j.tsf.2014.03.069.

[34] Chung Y, Park H, Baek S, Soo Y, Kim D. Processing Research Electrophoretic deposition behavior of ZnO nanoparticles and their properties on conductive fabrics. J Ceram Process Res 2014;15:331–5.

[35] Liu Y, Ata MS, Shi K, Zhu G-Z, Botton GA, Zhitomirsky I. Surface modification and cathodic electrophoretic deposition of ceramic materials and composites using celestine blue dye. RSC Adv 2014;4:29652. doi:10.1039/C4RA03938F.

[36] Rassaei L, Marken F, Sillanpää M, Amiri M, Cirtiu CM, Sillanpää M. Nanoparticles in electrochemical sensors for environmental monitoring. Anal Chem 2011;30:1704–15. doi:10.1016/j.trac.2011.05.009.

[37] Qiu Y, Li J, Li H, Zhao Q, Wang H, Fang H, et al. A facile and ultrasensitive photoelectrochemical sensor for copper ions using in-situ electrodeposition of cuprous oxide. Sensors Actuators B Chem 2015;208:485–90. doi:10.1016/j.snb.2014.11.061.

[38] Chung H-J, Sulkin MS, Kim J-S, Goudeseune C, Chao H-Y, Song JW, et al. Stretchable, multiplexed pH sensors with demonstrations on rabbit and human hearts undergoing ischemia. Adv Healthc Mater 2014;3:59–68. doi:10.1002/adhm.201300124.

[39] Narang J, Malhotra N, Singh S, Singh G, Pundir CS. Monitoring analgesic drug using sensing method based on nanocomposite. RSC Adv 2015;5:2396–404. doi:10.1039/C4RA11255E.

[40] Batra B, Lata S, Pundir CS. Construction of an improved amperometric acrylamide biosensor based on hemoglobin immobilized onto carboxylated multi-walled carbon nanotubes/iron oxide nanoparticles/chitosan composite film. Bioprocess Biosyst Eng 2013;36:1591–9. doi:10.1007/s00449-013-0931-5.

[41] Vernardou D, Drosos H, Spanakis E, Koudoumas E, Savvakis C, Katsarakis N. Electrochemical and photocatalytic properties of WO3 coatings grown at low temperatures. J Mater Chem 2011;21:513. doi:10.1039/c0jm02413a.

[42]   Yuan S-J, He H, Sheng G-P, Chen J-J, Tong Z-H, Cheng Y-Y, et al. A photometric high-throughput method for identification of electrochemically active bacteria using a WO3 nanocluster probe. Sci Rep 2013;3:1315. doi:10.1038/srep01315.

[43]   Wang C, Yin L, Zhang L, Xiang D, Gao R. Metal oxide gas sensors: Sensitivity and influencing factors. Sensors 2010;10:2088–106. doi:10.3390/s100302088.

[44]   Walia S, Balendhran S, Nili H, Zhuiykov S, Rosengarten G, Wang QH, et al. Transition metal oxides - Thermoelectric properties. Prog Mater Sci 2013;58:1443–89. doi:10.1016/j.pmatsci.2013.06.003.

[45]   Zheng H, Ou JZ, Strano MS, Kaner RB, Mitchell A, Kalantar-Zadeh K. Nanostructured Tungsten Oxide - Properties, Synthesis, and Applications. Adv Funct Mater 2011;21:2175–96. doi:10.1002/adfm.201002477.

[46]   More AJ, Patil RS, Dalavi DS, Mali SS, Hong CK, Gang MG, et al. Electrodeposition of nano-granular tungsten oxide thin films for smart window application. Mater Lett 2014;134:298–301. doi:10.1016/j.matlet.2014.07.059.

[47]   Zhu T, Chong MN, Seng E. Nanostructured Tungsten Trioxide Thin Films Synthesized for Photoelectrocatalytic Water Oxidation: A review. ChemSusChem 2014;7:2974–97. doi:10.1002/cssc.201402089.

[48]   Chen Z, Lv H, Zhu X, Li D, Zhang S, Chen X, et al. Electropolymerization of Aniline onto Anodic WO3 Film: An Approach to Extend Polyaniline Electroactivity Beyond pH 7. J Phys Chem C 2014;118:27449–58. doi:10.1021/jp509268t.

[49]   Choi Y, Kim S, Seong M, Yoo H, Choi J. Applied Surface Science NH4-doped anodic WO3 prepared through anodization and subsequent NH4OH treatment for water splitting. Appl Surf Sci 2015;324:414–8. doi:10.1016/j.apsusc.2014.10.059.

[50]   Liu L, Layani M, Yellinek S, Kamyshny A, Ling H, Lee PS, et al. "Nano to nano" electro-deposition of WO3 crystalline nanoparticles for electrochromic coatings. J Mater Chem A 2014;2:16224–9. doi:10.1039/C4TA03431G.

[51]   Rodríguez-Pérez M, Chacón C, Palacios-González E, Rodríguez-Gattorno G, Oskam G. Photoelectrochemical water oxidation at electrophoretically deposited WO3 films as a function of crystal structure and morphology. Electrochim Acta 2014;140:320–31. doi:10.1016/j.electacta.2014.03.022.

[52]   Zou B, Gong S, Wang Y, Liu X. Tungsten Oxide and Polyaniline Composite Fabricated by Surfactant-Templated Electrodeposition and Its Use in Supercapacitors. J Nanomater 2014;2014:1–9. doi:10.1155/2014/813120.

[53]   Martín AJ, Chaparro AM. Influence of Operation Parameters on the Response of a PEMFC with Electrodeposited Pt-WO3 Cathode. Fuel Cells 2014;14:742–9. doi:10.1002/fuce.201300242.

[54]   Gui Y, Blackwood DJ. Electrochromic Enhancement of WO3-TiO2 Composite Films Produced by Electrochemical Anodization. J Electrochem Soc 2014;161:191–201. doi:10.1149/2.0631414jes.

[55] Anik M. Effect of concentration gradient on the anodic behavior of tungsten. Corros
      Sci 2006;48:4158–73. doi:10.1016/j.corsci.2006.03.014.

[56] Anik M, Osseo-Asare K. Effect of pH on the Anodic Behavior of Tungsten. J Electro-
      chem Soc 2002;149:B224. doi:10.1149/1.1471544.

[57] Chai Y, Chin IK, Yam FK, Hassan Z. Formation of Tungsten Oxide Nanostructures
      Prepared in Hydrochloric Acid. J Electrochem Soc 2014;161:202–6. doi:
      10.1149/2.014405jes.

[58] Moshofsky B, Mokari T. Electrochromic active layers from ultrathin nanowires of
      tungsten oxide. J Mater Chem C 2014;2:3556. doi:10.1039/c4tc00093e.

[59] Qi C-X, Tan Z, Feng Z-H, Yu L-P. Fabrication of bowl-like porous WO3 film by col-
      loidal crystal template-assisted electrodeposition method. J Mater Sci Mater Electron
      2014;25:1553–8. doi:10.1007/s10854-014-1767-8.

[60] Shi Y, Liu Y, Xin Y. Study on the Properties of Photocatalytic Degradation of DMP on
      WO3/TNas Photoelectrodes. Adv Mater Res 2014;850-851:12–5. doi:10.4028/
      www.scientific.net/AMR.850-851.12.

[61] Ali H, Ismail N, Hegazy A, Mekewi M. A novel photoelectrode from TiO2-WO3
      nanoarrays grown on FTO for solar water splitting. Electrochim Acta 2014;150:314–9.
      doi:10.1016/j.electacta.2014.10.142.

[62] Dulgerbaki C, Uygun A. Efficient Electrochromic Materials Based on PEDOT/WO3
      Composites Synthesized in Ionic Liquid Media. Electroanalysis 2014;26:2501–12. doi:
      10.1002/elan.201400369.

[63] Xin Y, Gao M, Wang Y, Ma D. Photoelectrocatalytic degradation of 4-nonylphenol in
      water with WO3/TiO2 nanotube array photoelectrodes. Chem Eng J 2014;242:162–9.
      doi:10.1016/j.cej.2013.12.068.

[64] Qi H, Wolfe J, Wang D, Fan HJ, Fichou D, Chen Z. Triple-layered nanostructured
      WO3 photoanodes with enhanced photocurrent generation and superior stability for
      photoelectrochemical solar energy conversion. Nanoscale 2014;6:13457–62. doi:
      10.1039/c4nr03982c.

[65] Lai CW. Photocatalysis and photoelectrochemical properties of tungsten trioxide
      nanostructured films. Sci World J 2014;2014:843587. doi:10.1155/2014/843587.

[66] Chappanda KN, Smith YR, Rieth LW, Tathireddy P, Misra M, Mohanty SK. TiO2-
      WO3 Composite Nanotubes from Co-Sputtered Thin Films on Si Substrate for En-
      hanced Photoelectrochemical Water Splitting. J Electrochem Soc 2014;161:H431–7.
      doi:10.1149/2.110406jes.

[67] Yang T, Zhang Y, Cai Y, Tian H. Effect of processing parameters on anodic nanopo-
      rous tungsten oxide film structure and porosity for hydrogen detection. J Mater Res
      2014;29:166–74. doi:10.1557/jmr.2013.369.

[68] Ayouchi R, Bhattacharyya SR, Barrado JRR, Schwarz R. Tungsten trioxide nanostructured electrodes for organic dye sensitised solar cells. Int J Nanotechnol 2014;11:869–81.

[69] Baião P. Nanostructuring silicon probes via electrodeposition: Characterization of electrode coatings for acute in vivo neural recordings. MSc thesis. Universidade Nova de Lisboa, Portugal, 2014.

[70] Heidari EK, Zamani C, Marzbanrad E, Raissi B, Nazarpour S. WO3-based NO2 sensors fabricated through low frequency AC electrophoretic deposition. Sensors Actuators B Chem 2010;146:165–70. doi:10.1016/j.snb.2010.01.073.

[71] Bai S, Zhang K, Wang L, Sun J, Luo R, Li D, et al. Synthesis mechanism and gas-sensing application of nanosheet-assembled tungsten oxide microspheres. J Mater Chem A 2014;2:7927. doi:10.1039/c4ta00053f.

[72] Santos L, Silveira CM, Elangovan E, Neto JP, Nunes D, Pereira L, et al. Synthesis of WO3 Nanoparticles for biosensing applications. Submitted 2015.

[73] Wu Y, Hu S. Biosensors based on direct electron transfer in redox proteins. Microchim Acta 2007;159:1–17. doi:10.1007/s00604-007-0749-4.

[74] Feng J-J, Xu J-J, Chen H-Y. Direct electron transfer and electrocatalysis of hemoglobin adsorbed onto electrodeposited mesoporous tungsten oxide. Electrochem Commun 2006;8:77–82. doi:10.1016/j.elecom.2005.10.029.

[75] Zhou DD. Microelectrodes for in-vivo determination of pH. Electrochem. sensors, Biosens. their Biomed. Appl., Academic Press, Inc.; 2008, p. 261–305.

[76] Bandodkar AJ, Hung VWS, Jia W, Valdés-Ramírez G, Windmiller JR, Martinez AG, et al. Tattoo-based potentiometric ion-selective sensors for epidermal pH monitoring. Analyst 2013;138:123–8. doi:10.1039/c2an36422k.

[77] Cogan SF. Neural stimulation and recording electrodes. Annu Rev Biomed Eng 2008;10:275–309. doi:10.1146/annurev.bioeng.10.061807.160518.

[78] Meyer RD, Cogan SF, Nguyen TH, Rauh RD. Electrodeposited iridium oxide for neural stimulation and recording electrodes. IEEE Trans Neural Syst Rehabil Eng 2001;9:2–11.

[79] Venkatraman S, Hendricks J, King ZA, Sereno AJ, Richardson-Burns S, Martin D, et al. In vitro and in vivo evaluation of PEDOT microelectrodes for neural stimulation and recording. IEEE Trans Neural Syst Rehabil Eng 2011;19:307–16. doi:10.1109/TNSRE.2011.2109399.

[80] Conway BE, Pell WG. Double-layer and pseudocapacitance types of electrochemical capacitors and their applications to the development of hybrid devices. J Solid State Electrochem 2003;7:637–44. doi:10.1007/s10008-003-0395-7.

5

Electrodeposition of Thin Films for Low-Cost Solar Cells

Abhijit Ray

Abstract

This chapter endeavors to bring to the reader the potential of electrochemical deposition processes to develop low cost solar cells. The targeted photovoltaic absorbers are largely inorganic compound semiconductors containing earth abundant elements and satisfying the physical criteria of band gap and dopability. Electrodeposition, being an well established chemical process having its inherent capability for a large productions, should also be applied to develop various low cost solar cells, such as Cu_2ZnSnS_4, Cu_2S, FeS_2, Zn_3P_2, few/mono-layer MoS_2 and WS_2 etc having direct band gaps and ability to engineer their band gap through in-situ doping for creating various possible hereojunctions. The author believes that the combined effort of electrochemistry, physics and chemical engineering can push this simple and ultra-low cost technique of electrodeposition along with other supporting solution based techniques to produce PV energy at most affordable price in coming days.

Keywords: Solar cells, Electrodeposition, Thin films

1. Introduction

Despite the fact that, solar photovoltaic (PV) devices and modules made from crystalline (poly/mono) Silicon dominate the market at present, a significant efforts are underway in developing alternative materials to reduce dependence on Silicon [1,2] and bring them in the form of thin films to reduce the module cost [3,4]. Such non-silicon materials satisfying the prime requirement of right optical band gap (between 1-1.6eV) and high absorption coefficients (in the case of direct band gap semiconductors) [5] in the visible solar spectrum are usually binary or multinary alloys or compounds [6]. In most of the cases, their heterojunctions with a wide band gap emitter layer are exploited in the solar cells as their band gap can be tuned to a desired level by varying the compositions [7-9]. Because of the stoichiometry and stability of these materials are a primary concern to fabricate a successful junction [10,11], they are usually grown by highly controlled vacuum environments, such as the vacuum evaporation [12,13], chemical vapor deposition (CVD) [14-17], molecular beam epitaxy (MBE) [18], or sputtering [19-21] techniques. Besides the high energy requirement for the above film processing, emission of gaseous waste materials is another serious issue with these techniques [22,23]. These materials can also be deposited in thin films from solution in so called 'wet process' having the advantage of 'low energy' and 'low waste', provided the chemical environment allows a controlled ionization and subsequent reactions to form the compound film with required stoichiometry [24-26]. Among wet processes, spin coating [27], ink printing [28,29], chemical spray pyrolysis [30-32], successive ionic layer adsorption and reaction (SILAR) [33], chemical bath deposition (CBD) [34] and electrochemical deposition or electrodeposition (ED) [35,36]. In general, thin films deposited by ED method do not possess the high crystalline perfection as that made from CVD or MBEs. However, it has appeared as an attractive method owing to the involvement of cheap capital equipment [37-39], associated scalability [40,41], low waste of precursors and no usage of air or carrier gasses in most cases [36,42]. In addition, for application in PV cells, ED process allows one to easily alter both the optical properties (band gap), electronic properties (carrier concentration, conductivity and mobility) and structure (lattice constant) by the composition modulation through the control of bath parameters, such as applied potential (magnitude and type – dc or pulse), pH, and temperature of the bath, use of agitation (stirring) or excitation (such as by applying light). A number of largely cited ED processes for the synthesis of various low cost compound semiconductors for photovoltaic applications, their reported optical band gap and champion cell efficiencies are summarized in Table-1. Thus, the electrodeposition technique deserves its place for large scale commercialization of various new PV devices and modules of lower cost than Silicon counterpart in future. The present chapter brings to the reader's attention, the common aspects of electrodeposition technique and its application in some specific PV absorber materials.

System	Research group	Year	Method of electrodeposition	Obtained band gap (eV)	Best cell efficiency reported (%)	Reference
Cu_2S	Anwar *et al.*	2002	Cathodic	-	-	[155]
Cu_2SnS_3	Koike *et al.*	2012	Potentiostatic, single step	-	2.84	[156]
	Mathews *et al.*	2013	Potentiostatic, stacking metallic layers	1	-	[157]
Cu_2ZnSnS_4	Mkawi *et al.*	2014	Pulsed electrodeposition	1.36-1.47	1.66	[172]
	Araki *et al.*	2009	Potentiostatic, single step	-	0.98	[92]
	Ennaoui *et al.*	2009	Potentiostatic, single step	1.54	3.4	[36]
	Sheng *et al.*	2012	Potentiostatic, single step using ionic liquid	1.51	-	[158]
	Ahmed *et al.*	2012	Potentiostatic, stacking mettalic layers	1.48-1.6	7.3	[130]

System	Research group	Year	Method of electrodeposition	Obtained band gap (eV)	Best cell efficiency reported (%)	Reference
	Cui *et al.*	2011	Potentiostatic, single step	1.5	-	[159]
	Pawar *et al.*	2011	Potentiostatic, single step	1.48-1.76	-	[160]
	Chan *et al.*	2010	Potentiostatic, single step using ionic liquid	1.49-1.5	-	[154]
$Cu_2ZnSnSe_2$	Jeon *et al.*	2014	Potentiostatic, single step	1.04	8	[161]
	Zhang *et al.*	2013	Potentiostatic, single step	0.98	4.5	[129]
	Guo *et al.*	2012	Potentiostatic, stacking mettalic layers	1.29	7	[162]
	Septina *et al.*	2013	Potentiostatic, single step	1.41-1.48	-	[163]
	Li *et al.*	2012	Potentiostatic, single step	1	1.7	[164]
SnS	Seichen *et al.*	2013	Cathodic	-	-	[165]
	Ghazali *et al.*	1998	Cathodic	1.1	-	[166]
	Subramanian *et al.*	2001	Cathodic	1.15	-	[167]
	Zainal *et al.*	1996	Cathodic	-	-	[168]
	Cheng *et al.*	2006	Galvanostatic (constant current)	1.21-1.42	-	[169]
	Cheng *et al.*	2007	Pulsed electrodeposition	1.23-1.33	-	[170]
	Cheng *et al.*	2006	Cathodic	1.48-1.24	-	[171]

Table 1. Various reported electrodeposition (ED) routes for fabrication of some low cost photovoltaic grade thin films and their associated properties. Although several other systems including $CuIn(S/Se)_2$, $Cu(In,Ga)(S/Se)_2$, GaAs etc have been attempted via electrodeposition, they are not listed below as they either involve expensive precursors or post treatment.

2. Electrodeposition

Electrodeposition of metals and alloys on a conducting substrate involves the reduction of metal ions from various possible electrolytes, such as aqueous, non-aqueous, organic, ionic liquids and molten salts. In general the reduction of a cation species present in the aqueous solution with z positive charge, M_z^+ to produce a neutral metal atom M on the cathode as substrate is represented by,

$$M_z^+ + ze\left[solution\right] \rightarrow M\left[substrate\right] \tag{1}$$

In above reaction, the charges are provided by an external current source and a layer of M atoms are deposited in an allowed time. More is the concentration of M_z^+, larger will be the number of M atoms deposited onto the substrate to form a lattice eventually.

2.1. Common aspects

In the reaction given by Eq. 1, the process involves charged species at the interface between a solid conducting electrode and a liquid solution. These charged species are obviously the cations and electrons, which cross the interface. There are usually four types of primary characteristics associated to Eq. 1 which are commonly dealt in the case of electrodeposition [43]: 1. *Interface between metal and solution.* Here crystal structure and electronic properties of the metal, molecular structure of the polar solvent (in case of aqueous medium), and the structure and properties of the electrolyte used are the governing factors. 2. *Rate of the deposition process.* This is proportional to the concentration of the active cations in the solution available for reduction by the electrons supplied by the cathode. A proportionality constant can be referred to as 'rate constant', k that is defined in terms of an electrochemical activation energy, ΔG_e which in turn is determined as a function of the electrode potential, E through following set of equations:

$$k = \frac{k_B T}{h} \, ln\left(-\frac{\Delta G_e}{RT} \right) \tag{2}$$

$$\Delta G_e = f\left(E\right) \tag{3}$$

3. *Nucleation and growth mechanism.* As the ions reach the electrode surface for reduction the surface nuclei can grow in one, two or three dimension as determined by the parameters of the electrodeposition and nature of the solution. This may lead to the formation of monolayer, few layers, multilayers or three dimensional growths depending on the mode of nucleation. 4. *Structure and physico-chemical properties of the deposit.* Crystal structure, microstructure and elementary composition of the deposit under a certain growth condition will determine the physical (such as, electrical, mechanical, optical etc.) as well as chemical (such as, reactivity to its ambient, stability etc.) properties of the film which is desired for an application or required to be looked at to prevent certain unwanted effect during an application. A more detail of nucleation and growth mechanism is given in reference [43] and in the subsequent section in brief.

2.2. Nucleation and growth mechanisms

The electrodeposition process attains a dynamic equilibrium among the metal cations, electrons participating in the reduction process and the metal atoms to be deposited on the cathode surface. With the establishment of this dynamic equilibrium, fully solvated (water molecules attached to it in case of an aqueous electrolyte or ligands in the case of use of complexing agents) ions get attracted towards the oppositely charged electrode surface by coulombic forces as shown in Fig. 1 in the case of cations, for example. Eventually an electrical double layer is formed, where the inner layer is largely composed of oriented water dipoles interposed by some preferentially adsorbed ions and the outer layer is a jacket of ions having charge opposite to that of the electrode. These cations are adsorbed onto the cathode surface by displacing the attached water molecules or ligands. Upon a continuous depletion of the depositing ions from the double layer region, fresh ions are supplied from the bulk of electrolytes by either or mix of three important processes: (i) diffusion due to concentration gradient, (ii) electric field assisted migration and (iii) convection current in the electrolyte due to temperature or agitations.

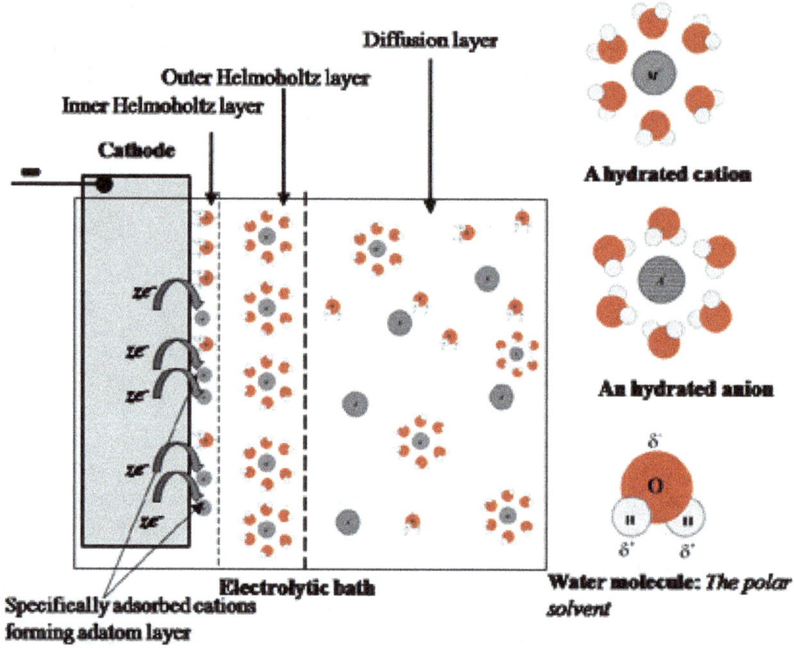

Figure 1. Illustration of a fundamental cathodic electrodeposition process at for a single constituent metal salt dissolved in an aqueous electrolyte

Some of the process parameters critically determine the growth of the deposits, such as current density through the electrode, electrolytic bath condition, shape of the active electrode and type and shape of the counter electrode. The current density determines in particular, the rate of deposition and hence microstructure the deposited film. An optimum range of current density is determined experimentally for each of the depositing ion species by electrochemical characterizations technique. Bath parameters, such as the concentration of individual ion species, use of complexing agents and their concentration, temperature, illumination or irradiation influences the nucleation process as these parameters can directly influence the inner row (Stern layer) of the Helmoholtz double layer. Bath temperature controls the rate of diffusion of the ions from the diffusion to Helmohltz layer, the convection current, stability of any complex and decomposition of any additive used. The shape of the active electrode sometimes determines the thickness uniformity of the deposits. Higher current density at the edges or projections as compared to the crevices and hollow spaces leads to the formation of thicker growth at the edges. The counter electrode is used to complete the circuit and balance the charge in dynamic process during electrodeposition. As the number of anions and OH^- remains very high as compared to the positive ions and protons as in the case of cathodic deposition, the counter electrode area becomes an important factor. Usually, a counter electrode area should be as large as four times the geometric surface area of the active electrode.

2.3. Cathodic electrodeposition

Cathodic electrodeposition, that means the deposition of thin film on the cathode surface is especially important for the single as well as multilayered metals or alloys deposition for compound semiconductor thin films for solar cells, photoelectrochemical (PEC) cells and photocatalytic action in Hydrogen Evolution Reaction (HER) in the case of PEC water splitting. The electrodeposition of an alloy requires their ions must be present in an electrolyte where the individual deposition potentials can be made to be close or even the same at the best.

There are three main steps must be followed in a cathodic electrodepositon process:

1. *Ionic migration:* The hydrated ion(s) in the electrolyte migrate(s) toward the cathode under the influence of an applied potential as well as through diffusion and/or convection.

2. *Electron transfer:* At the cathode surface the hydrated metal ion(s) enter(s) the diffusion (double) layer where the water molecules of the hydrated ion are aligned by the weak field present in this layer. Eventually, the metal ion(s) enter(s) the fixed Helmoholtz layer, where, due to the higher field present, the hydrated shell is lost. Subsequently, the individual ion is neutralized by electron transfer from cathode and is adsorbed at the cathode surface.

3. *Incorporation:* In this last step the adsorbed atom is drifted to a growth point on the cathode and is incorporated in the growing lattice. The process continues, largely, as long as the applied potential is near to the reduction potential(s) of the ion(s).

The replenishment of the metal ions M_z^+ can be possible in two ways: 1. M_z^+ *from an anode of* M: the metal to be deposited is used as anode, when a metal ion is produced from some surface atoms in the metal lattice leaving back the electron(s) to anode. 2. M_z^+ *from a salt of* M: the anode can be made of a metal or conducting electrode other than M, a *Pt* electrode in most of the cases.

At the same time, ions from the solution are deposited on the electrode, according to step 2 above. The specific potentials at which these two reaction rates are equal, called *standard potentials,* which are defined for solutions maintained at 25°C and at an activity value of unity. Again, as the above two processes (oxidation-reduction) are reversible, the equilibrium potential at the metal electrode is given by the well known *Nernst expression:*

$$E = E^0 + \frac{RT}{zF}\ln Q \qquad (4)$$

Here, E^0 is the *standard oxidation/reduction potential* for the electrode immersed in a solution of ions having unit activity, R is the gas constant, T is the temperature, z represents the number of valence electrons participating in the redox process, F is the Faraday constant ($F = 9.648\,533\,99(24){\times}10^4\ C\ mol^{-1}$) and Q is the reaction quotient that may be defined as the ratio of concentration of oxidized to that of the reduced ion species, that is usually taken as the activity of the reactant ions (a). However, the deposition, as a rule is an irreversible process that has to occur at some extra potential, E_{over} (overpotential) required for maintaining the reduction reaction and therefore, Eq. (4) should be written as,

$$E_{dep} = E^0 + \frac{RT}{zF}\ln a^{z+} + E_{over} \qquad (5)$$

In practice, the metal deposition potential, E_{dep} can be estimated easily by cyclic voltammetry technique by knowing the values of a^{z+} and E_{over} for a fixed plating condition, including bath parameters, such as the current density and temperature, as well as the ionic parameters, such as solution pH, concentration, valence, and mobility.

In some cases of alloy deposition, when the standard electrode potential of more than one types of ions are widely different, such as in the case of $Zn^{2+}{\rightarrow}Zn$ and $Cu^{2+}{\rightarrow}Cu$, $E_0^{Zn2+;Zn} = -0.762V$ and $E_0^{Cu2+;Cu} = -0.345V$, respectively, their co-electrodeposition is apparently impossible. However, the difference can be eliminated (or reduced) by changing the values of the activities. This can be achieved by introducing a considerable change in ionic concentrations via complex ion formation.

2.3.1. Factors affecting the cathodic electrodeposition

In the case of aqueous electrolytes used in the electrodeposition, hydrogen evolution at the cathode is a common problem as the hydrogen evolution reaction (reduction of a proton released from water molecule by an electron) is pH dependent and it can occur simultaneously with the adsorption of cations and their reduction by electrons. This may cause pitting of the deposited thin film and worsening of its morphology. In order to prevent this hydrogen accumulation, a suitable wetting agent may be used to promote detachment of the H_2 evolved at the cathode. Sometimes by the use of ionic liquid or non-aqueous electrolyte, the H_2 evolution can be prevented. The control of pH in electrodeposition is important. The pH controls the overall conductivity of the electrolyte. If the electrolyte by itself is not very conductive, a suitable acid or alkali or salt may be used to achieve required conductivity. However, the pH is required to be optimized for an electrodepositon, because a low pH of solution may result only H_2 evolution, whereas a high pH may cause the inclusion of –OH group in the deposits.

Current density plays an important role on the deposits. Use of low current densities, in general, result in higher impurities in the deposit, which in turn affects residual stress and other properties of the deposit. The flatness of the cathode surface (substrate used) also plays an important role in the case of alloy deposition, in general. An uneven surface leads to the variation of current densities and hence, the throwing power. In the case of inferior throwing power, the result may be uneven alloy composition on the cathode.

2.4. Anodization

The anodization is an important age-old electrochemical technique that has long been used for increasing the corrosion and wear resistance of various metal surfaces. It is an electrolytic surface passivation technique used to increase the thickness of native oxide layer on the surface of metal under the application of a large bias. Usually few tens of volts are required to maintain the oxidation reaction on the surface of anode. Eventually thickness of the oxide layer increases. Anodization changes the preferential crystal orientation (microscopic texture) of the surface and sometimes changes the crystal structure of the metal near the surface. The anodized surface retains its mechanical strengths and it can sustain mechanical deformation easily. Depending on the applied potential and anodization time, it may be referred to as 'soft' or 'hard' anodization. In the case of soft anodization, the formed layers are thin and usually compact, whereas in the case of hard anodization, they are thicker and many a times porous or template structure. While the former has certain importance in hardening the metal surface, the latter is useful in producing nano-structure templates in the case of many functional thin films including the thin film nanostructured solar cells.

2.4.1. Basics of anodization processes

A standard anodization procedure includes three important steps: 1. alkaline cleaning, 2. acid activation, and 3. biasing and reaction. After cleaning the anode surface (where film is required), acid treatment is given to the surface in a mixture of nitric acid (HNO_3) and hydrofluoric acid (HF) to remove the native metal oxide layer and surface contaminants. The biasing and anode reaction are carried out in an electrochemical bath, which usually has a three-electrode configuration (metal anode, platinum cathode and Ag/AgCl reference electrode). When a constant voltage (potentiaostatic condition) or current (glavanostatic condition) is applied between the anode and cathode, electrode redox reactions along with the field-driven ion diffusion lead to the formation of an oxide layer on the anode surface. The redox reactions for anodizing of a metal M with z nos of valence electrons therefore can be expressed as,

$$\text{At the } M \,/\, M_2O_z \text{ interface}: \text{M} \leftrightarrow M^{z+} + ze^- \tag{6}$$

At the Ti oxide/electrolyte interface:

$$zH_2O \leftrightarrow zO^{2-}zH^+ \tag{7}$$

$$2H_2O \leftrightarrow O_2 + 4H^+ + 4e^- \tag{8}$$

At both interfaces:

$$2M^{z+} + zO^{2-} \leftrightarrow M_2O_z + ze^- \tag{9}$$

Some time due to the higher resistivity of the metal oxide film than the electrolyte and the metallic substrate, the deposition current density drops over time as the oxide film grows on the anode. As long as the electrical field is strong enough to drive the ion conduction through the oxide, the film keeps growing. For this reason, a large bias (even larger than the water splitting potential) is maintained. Obviously, the final oxide thickness, d is almost linearly dependent on the applied voltage, V_{app} according to:

$$D \sim aV_{app} \tag{10}$$

Where a is usually a constant lying the range 1.5-3 nm/V.

The physical and chemical properties (such as degree of nanometer roughness, morphology, stoichiometry, etc.) of the anodized film varies over a wide range according to different process parameters, such as applied potential (voltage), current density, electrolyte composition, pH, and temperature. If the applied voltage exceeds the dielectric breakdown limit (DBL) of the oxide, it will no longer be resistive to prevent further current flow. Eventually, it leads to more gas revolution and sparking, resulting degradation of film quality. However, the DBL varies in various electrolytes [44]. In their study, it was shown that, below the DBL, the anodic oxide film is thin and usually non-porous using non-fluorine electrolytes. A constant temperature during the anodization process is usually preferred to maintain a homogeneous field-enhanced dissolution over the entire area of anode. Since increased temperature accelerates the chemical dissolution rate, the working temperature is often kept relatively low to prevent the oxide from totally dissolving. Nature of the electrolyte has been found to have great effect as well. Sul *et al.* [45] have shown highest thickness of anodized TiO_2 film is obtained when H_2SO_4 is used.

2.4.2. Anodization used for solar cell fabrication

Anodization has reportedly been applied in various aspects of surface nano-engieering in solar cells and PEC cells through anodized TiO_2 nanostructure, anodic alumina (AAO) template etc. TiO_2 is a very suitable oxide material for dye-sensitized solar cells (DSSC), because of its extraordinary oxidizing ability of photogenerated holes. TiO_2 thin films are prepared by various preparation methods, but the efficiency of the DSSC is strongly enhanced by the increased dye absorption capacity of the photoelectrode and its nanostructure has been found to play a beneficial role [46,47]. Researchers have successfully adopted the anodization technique to grow the nanotube arrays from thin Ti films as well using anodization on a variety of substrates, such as glasses [58-60], conducting glass (FTO) [51-53], and silicon [54-56]. Another application of anodization is the removal through dissolution to produce rough surfaces, which are sometime very useful in creating anti-reflection coating as well as nano-structure to enhance optical path through scattering for solar cells. In this direction, a significant improvement of light management for a:Si solar cell was reported by Huang *et al.* [57] where they reported the use of AAO template to deposit Ag nano-back reflector to improve optical path in the system. It was reported to improve the photo conversion efficiency from 6.64% to 7.11% in a:Si solar cells. Wang *et al.* [58] used AAO template to produce nano-wire of Cu_2ZnSnS_4 which is a direct band gap photovoltaic absorber. Anodized ZnO post doped with Al has proved to be an excellent textured transparent conducting oxide (TCO) layer for thin film silicon solar cells [59].

2.4.3. Anodic vs. Cathodic Electrodepositions of compound semiconductors for solar or PEC cells

In principle, both anodic and cathodic deposition may be used for the fabrication of compound semiconductor for thin film solar cells. However, there are two issues with anodic deposition. Firstly the stiochiometry of the compound may change over time of deposition, as the anions available from the solution are only available to participate in the redox reaction at the metal anode, such as in the case of CdS, CdTe, ZnTe etc. Secondly, if more than one metal components are required in the film, such as the ternary or quaternary compounds like $CuIn(S/Se)_2$, $Cu_2ZnSn(S/Se)_2$ etc anodic deposition is not much suitable and cathodic deposition can handle all the cationic components in solution easily provided their reduction potential can be made close. It has been reported that CdS can be made by the anodic deposition of sulphur from a solution containing S^{2-} ions on a cadmium anode [60]. Similar anodic deposition has been attempted for CdTe films [61] as well, however, the stoichiometry of the deposit was not easy to regulate. However, if each element of a semiconducting material (e.g. Cd and Te for CdTe, or Cd and S for CdS) can be deposited cathodically, it may be possible to deposit them in good stoichiometry. Cathodic deposition can occur either by deposition of the individual cation and anion components in the required ratio in a multi-step process, or by co-deposition resulting from mass transfer and decomposition of a complex containing the cation and anion components.

2.5. Pourbaix diagram and its application in electrodeposition

Prior to conducting an electrodeposition, the analysis of thermodynamic data obeying chemical and electrochemical equilibrium is highly helpful in understanding the reactivity of a compound system to be used for deposition of a desired phase. *Pourbaix* (potential–pH) diagrams [62] offers an excellent guideline about the detailed picture of the electrochemical solution growth system in terms of the deposition variables and reaction possibilities under different conditions of pH, redox potential, and/or concentrations of dissolved and electroactive species.

Pourbaix diagrams for a number of solar energy materials (such as, that for the aqueous Cd–S, Cd–Te, Cd–Se, Cu–In–Se, and Sb–S systems) have been presented and discussed in detail by Savadogo [63]. Pourbaix diagram is even more helpful in case, the system is partially substituted for an element by same group element for the purpose of doping. Dremlyuzhenko et al. [64] analyzed theoretically the mechanisms of redox reactions in the $Cd_{1-x}Mn_xTe$ and

$Cd_{1-x}Zn_xTe$ aqueous systems and demonstrated the variation of physicochemical properties of the semiconductor surfaces as a function of solution pH. Pourbaix diagrams regarding compound semiconductors can be found from various literatures in the field of geochemistry, mineral processing, surface treatment and corrosion sciences. A number of important diagrams related to solar energy materials are reproduced in the reviews by Savadogo [63] and the monograph by Bouroushian [65].

2.6. Electrochemical epitaxy

The ordered growth of a single crystalline material on the top of a crystalline substrate is known as epitaxy, one of the important processes in semiconductor technology wherever crystalline thin films are required. Some of the most common epitaxial techniques are molecular beam epitaxy (MBE) [66], metal organic vapor phase epitaxy (MOVPE) [67] and liquid phase epitaxy (LPE) [68]. Among these techniques, the electrochemical atomic layer epitaxy (ECALE) is an LPE technique having two specific advantages over vacuum based epitaxy techniques, such as MBE and MOVPEs: 1. reduced cost of materials usage and operation, 2. ECALE is performed at room temperature or under the boiling point of the solvent, which are treated as "low temperature" deposition. The ECALE method is based on the alternate underpotential deposition (UPD) of the elements on a single crystal conducting substrate in a cycle [69]. The UPD is a surface-limited phenomenon, enabling the deposit to grow in atomic layers. Simply, the UPD is the one condition of deposition where an atomic layer of a first element is deposited on a second, at a potential prior to that needed to deposit the first element on itself. Thus each deposition cycle forms a monolayer of the compound, and the thickness of the deposit can be controlled by the number of deposition cycles. UPD is a thermodynamic phenomenon, where the interaction energy between the two elements is larger than that of an element with itself. Due to its formation on a single crystal surface, the resulting thin film consists of surface compound or alloy having same crystalline nature as that of the substrate.

In illustrative ECALE technique for form epitaxial CdS on (111) oriented Ag substrate (A detail of experimental conditions are described in ref. Innocenti, M., et al. [70]) is shown in Fig 2. The cycle is composed of four steps: reductive UPD of cadmium from a Cd^{2+} ion solution, a blank rinse, oxidative UPD of sulfur from a S^{2-} ion solution, and a second blank rinse. Separate solutions are used for each reactant and different potentials for each cycle step. The use of separate solutions and potentials provides extensive control over deposit growth, composition, and morphology. An atomic layer of S is deposited on one of Cd, and one of Cd is deposited on one of S according to following reactions:

$$\overset{\textit{Under UPD of Sulfur oxidation}}{S^{2-} \quad \rightarrow \quad S + 2e^-} \tag{11}$$

$$\overset{\textit{Under UPD of Cd reduction}}{Cd^{2+} + 2e^- \quad \rightarrow \quad Cd} \tag{12}$$

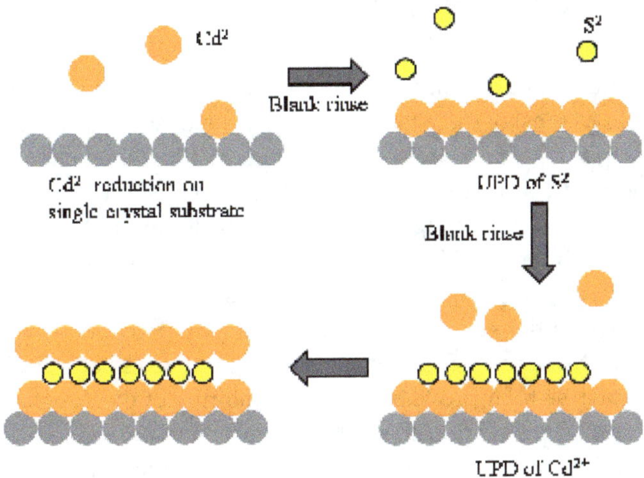

Figure 2. Illustration of an ECALE mechanism for deposition of CdTe thin film

A handful number of epi-films of compound semiconductors and solar energy materials have been attempted by various groups. Nanofilms of II-VI including ZnSe [71], CdTe [72,73], CdS [74-76], III-V compounds, such as InAs [77,78], and the photovoltaic absorber materials, such as CdTe [79], Ge, CIS, and CIGS.

3. Low cost thin film solar cells

Solar photovoltaics is projected to make a significant share of the world electricity production as the PV capacity is expected to grow up to 4.6TW worldwide by the year 2050 [80]. Inorganic thin-film semiconductors photovoltaic devices with a thickness of about 1 μm and efficiency of about 10% requires approximately, 10 g of active material per square meter, which corresponds to approximately 10 metric tons of the materials per GW production [6]. This estimate accounts to the requirement of an enormously large quantity of materials if a substantial share by thin films is to be considered, even though the material requirement is negligibly small as compared to bulk Silicon counterpart. Out of the presently commercially available thin-film materials, CdTe and Cu(In,Ga)(Se,S)$_2$ or CIGS, the elements Te, In, and Se are not abundant. Beside this, Indium used in Indium Tin Oxide (ITO) also has a large demand as the best performing transparent conducting electrode in many display devices. Therefore, a thin film photovoltaic technology beside their physical requirements (such as the optical band gap, absorption coefficient, intrinsic doping etc.) really demands its constituents to be abundant in earth crust (the relative abundance with respect to Silicon may be considered according to ref. [81]) as well as such elements have less superposition to other applications. By looking at the above aspects, a group of chalcogenides (CuS, Cu$_2$S, Cu$_2$SnS$_3$, Cu$_2$ZnSnS$_4$, SnS, FeS$_2$, MoS$_2$ and WS$_2$), oxides (CuO), silicides (β-FeSi$_2$, β-BaSi$_2$, Ca$_2$Si) and phosphides (CuP$_2$, Zn$_3$P$_2$, b-ZnP$_2$, SiP) can be identified as potential thin film photovoltaic absorbers [82].

4. Electrodeposited thin film solar cells

The categories of semiconducting compounds listed in the previous section can be considered for thin film solar cells as most of them (such as, Cu_2SnS_3, Cu_2ZnSnS_4, SnS, FeS_2, Zn_3P_2 etc) have direct optical band gap and high absorption coefficients in the visible solar spectrum. However, for these compound semiconductors after synthesis to be useful as PV absorber, stringent requirements are to be met regarding structure, composition, and morphology. Although, simple by operation, it is a formidable challenge for any electrodeposition process to take care of all the above mentioned requirements simultaneously. In many cases, more than a single step deposition and/or additional post deposition treatments (surface treatment, annealing etc) are required before electrodeposited layers can be deployed in a solar cell. The post deposition annealing has a special role to improve both the crystallinity of the deposits and the optoelectronic properties of the layer. In order to electrodeposit a compound semiconductor thin film with p-type conductivity and grain size of approximately a micrometer directly, a number of criteria have to be fulfilled. The nucleation density should be low to moderate so that large grains can be grown uniformly, and any secondary nucleation should be avoided. At the same time, overlapping grains should not leave pinholes that lower the shunt resistance of the device, causing a very low short circuit current density of the final cell. The layer thickness should be uniform throughout the substrate to ensure low surface recombination losses, higher average generation of electron-hole pairs in order to produce higher open circuit voltage.

As mentioned above, the requirement for strict control of stoichiometry has three important trade-offs: **1.** Electrons need to be readily available for the reduction of precursor ionic species at cathode (in the case of cathodic deposition) Again, free holes produced from photo- or thermal-excitation at the p-type layer being grown could reduce their numbers. **2.** Deposited atoms have to be incorporated into the correct positions in the growing semiconductor lattice. However, they may not take the lowest energy positions on the surface. In the case the atoms are deposited at higher free energy sites, the thermal energy may be insufficient to move them to a minimum free energy positions. **3.** The deposition rates and potential for the atomic constituents are required to be controlled so that the correct atoms are deposited in the correct order. For multinary system where this parameter is widely different, it becomes a subtle situation for the bath. The best approach to deal above issues of deposition has been attempted through ECALE as mentioned in the section 3.6. From the above discussion, it is therefore clear that, stringent control of ED parameters as well as post annealing could be a useful approach to produce p-type semiconductor thin film PV absorber layers.

4.1. Common procedures in electrodeposition of thin film solar cells

The electrodeposition followed by annealing (EDA) is a common procedure to for preparing good quality p-type compound semiconductors for solar cells. The primary purpose of the EDA is to form a reactive precursor film that can be converted to a compound semiconductor with the desired optoelectronic properties after a reactive annealing. For example, in the case where a compound chalcogenide film is required to be formed, the first step is to form an alloy layer or a stack of metal layers by ED, followed by annealing in a chalcogen atmosphere to form the desired semiconductor compound. As another example, in the case where a stoichiometric semiconductor film is required, such as CdTe, the alloy itself is produced in ED step and air annealing may be used to control the conductivity type or to improve crystalline nature (such as by $CdCl_2$ anneal). The real challenge appears when ED is applied to deposit thin films at larger area such as a PV module. Maintaining desired stoichiometry and compositional uniformity in whole area requires additional skill and control. The nature of deposited layers depends on the film substrate, precursor species in solution, applied potentiodynamic scheme, mass-transfer condition and bath temperature. A brief description of each of these parameters is given below:

i.   Nature of the substrate:

The substrate must be thoroughly cleaned, free of passivating oxides (if reactive to open air), and able to sustain the annealing conditions (should not soften or melt in post annealing). The substrate needs to be highly conducting to minimize the potential drop. Electrical contact to the substrate with electrode must not introduce any large resistance and also this contact must be sealed or protected from the solution environment.

ii.   Use of precursors:

In many cases, such as in the chalcogenide compounds the films contain from two to five elements. There, the metal precursor ions may be available in one or more oxidation states, each of which has different solubility and reduction potentials. The metal ion can be introduced from their respective salt in case a common bath is used or supplied by anodic dissolution of a sacrificial anode in the case of sequential deposition. The metal ions can also be complexed with ligands to increase their solubility, prevent precipitation, and more importantly to narrow down the reduction potentials of various metal ions. Other components of the ED bath are supporting electrolytes which are inert during ED process but increase conductivity of the solution, acids, bases or pH buffers to control the solubility of the ionic species, and organic additives to control reaction rates or to improve the morphology of the deposits.

iii.   Potentiodynamic condition:

Usually, different species of ion in solution has a different reduction potential. Theoretically, all the cations species can be reduced if the applied potential is made to a largest negative value to that of the ion species having largest negative reduction potential. However, such a deposit will be more likely in the form of binary and/or ternary compounds depending on the number of species in solution. Sometimes it leads to overgrowth or dendrite (fractal) formation of some elemental species (having lower reduction potential). Rather, by keeping the applied potential dynamic (potentiodynamic ED), this problem could be solved to some extent. It has been shown that use of a voltage sweep can assist the incorporation of electronegative elements and reduce any dendritic growth [83]. The work by Korger [84] suggest that, in a mixed ionic species system, the reduction potential of a more electronegative element in the compound may be shifted to positive, thermodynamically and a under-potential deposition condition may be achieved. Another common problem encountered during electrodeposition of ternary or quaternary chalcogen compounds is the presence of excess of the most noble metal (such as

Cu) in the film. This problem can be tackled either by reducing the concentration of the noble metal component to reduce its flux to the cathode surface, or by using an organic ligand to make complex of the metal ion in order to shift its reduction potential to more negative values.

iv.   Mass transfer condition:

As different ion species are to be reduced at the electrode has different sizes, their transport in the solution, from diffusion to double layer region and from the double layer to be adsorbed on cathode after electron transfer is mass transfer limited process (kinetics depend on their mass directly). Therefore, their horizontal, vertical or convective (agitation in the bath or in Rotating Disk Electrode) motion will have direct impact on the final deposit.

v.   Bath temperature:

An elevated bath temperature plays a key role to enhance surface diffusion of adatoms on the cathode surface in a cathodic ED. Due to enhanced adatom mobility on the surface, the crsystalline quality of the film can be improved. However, the temperature must be optimized first as it may adversely affect other components, such as the metal-complex stability, adhesion of the film etc.

4.2. Review of electrodepositied-annealed (EDA) thin film PV absorbers

In the EDA scheme as defined in the previous section for the electrodeposition of p-type photovoltaic absorber thin films, four possible scheme of depositions are found in most of the literatures: (i) sequential deposition of metal layers (a stack), (ii) co-deposition of an alloy (to be post treated to form desired compound), (iii) stacks of binary and/or ternary compounds, and (iv) co-deposition of the compound as a whole as shown in Fig. 3.

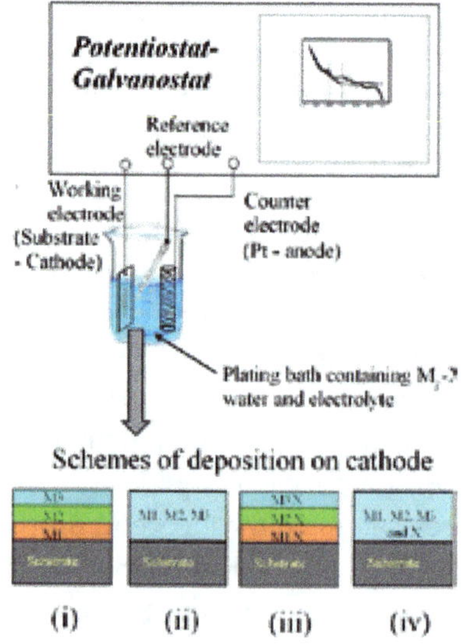

Figure 3. Schemes of cathodic electrodeposition of precursor film to be post annealed (EDA) for *p*-type photovoltaic absorber layers in literatures. Here, M_i represent constituent metal and X represents oxidized anion.

In the case (i) and (ii) above, the compound formation is done in post annealing step, when the anionic part (such as chalcogen for sulfides, selenides or tellurides) reacts with the metal precursors at a higher temperature, and the semiconductor layer grows usually with larger grains [85]. The limitations of this method are poor adhesion, inclusion of cracks, blisters, and pinholes resulting from the substantial volume increase during the solid state reaction. Deposition of CIGS [86, 87] and CZTS [88-92] has been attempted with these two schemes.

The deposition of binary compounds in two/more layers in scheme (iii) is a mass transport limited process which may become complicated by the formation of several phases with different stoichiometries within a narrow range of experimental conditions. The binary stacks react together during annealing step, make phase transformation to produce large grains of the ternary/quaternary semiconductor compounds. One advantage of this method is that there are no large changes in volume during the formation of the ternary semiconductor. The layer of $CuInSe_2$ has been found to form with good optoelectronic properties in the case of EDA involving a Cu-Se and In-Se compound layers [83, 93].

Scheme (iv) of ED is performed in a single bath containing all cationic and anionic precursors and it has been attempted for all major classes of low cost PV absorbers including. CdTe [94,95], CIS [96,97], and CIGS [98]. This scheme of ED is much slower due to the lower concentrations of constituent used in solution. The process is difficult to control in the case of ternary or quaternary chalcogenides, however the subsequent annealing step in the presence of excess chalcogen may be used to further react secondary phases. Sometime, majority of secondary phases appear at the surface which are required to be removed by further surface treatment, such as through selective chemical etching.

A brief account of ED and EDA for majority of p-type photovoltaic absorbers, material-wise is given in following sub-sections.

4.2.1. Electrodeposition of CdTe

Electrodeposition of CdTe was attempted long back by Panicker et $al.$ [74] where an aqueous solution containing excess of Cd^{2+} ions and TeO_2 was dissolved in acidic solution as $HTeO_2^+$. Ni and Sb:SnO2 transparent conducting glass were used as depositing electrode. The material was found to be n-type if pure solutions were used, whereas the presence of low concentration of copper impurities would produce p-type CdTe.

Thermodynamic aspects of cathodic electrodeposition have been discussed by Kroger [84], where the deposition potential of elemental Te and Cd can be obtained from the equilibrium potentials according to following reactions with accompanied Nernst equations:

Reaction	Corresponding deposition potential (vs. RHE)
$HTeO_2^+ + 3H^+ + 4e^- = Te + H_2O$	$E = 0.559 + 0.0148\, log[HTeO_2^+] - 0.0443\, pH$
$Cd_2^+ + 2e^- = Cd$	$E = -0.4025 + 0.0295\, log[Cd^{2+}]$

Reduction of Cd^{2+} needs a cathodic potential, again Te can be reduced at more cathodic potentials to form H_2Te as per following reaction:

Reaction	Corresponding deposition potential (vs. RHE)
$Te + 2H^+ + 2e^- = H_2Te$	$E = -0.740 - 0.0295\ log[H_2Te] - 0.059\ pH$

Finally, CdTe is deposited at a less cathodic potential according to:

Reaction	Corresponding deposition potential (vs. RHE)
$Te + Cd^{2+} + 2e^- = CdTe$	$E = 0.740 - 0.0295\ log[Cd^{2+}]$

At more cathodic potentials, CdTe can be reduced back to Cd and H_2Te according to:

Reaction	Corresponding deposition potential (vs. RHE)
$CdTe + 2H^+ + 2e^- = Cd + H_2Te$	$E = -1.217 - 0.0295\ log[H_2Te] - 0.059\ pH$

Therefore, the choice of potential is stringent by the regions in which CdTe is stable to both anodic and cathodic decomposition. In other word the CdTe deposition window is limited by the anodic limit of CdTe stability and by the cathodic deposition potential of bulk Cd.

Later, Basol *et al.* [99] had reported a CdS | CdTe heterojunction solar cell with 9.35% efficiency. They fabricated it by an EDA route, first electrodeposition of a CdS film onto indium tin oxide (ITO) coated glass followed by an electrodeposition of the CdTe layer The layered structure was annealed at 400° C to finalize the heterojunction. CdTe deposition on TCO glasses was also studied by Rajeshwar *et al.* [100-103]. Later, its large-scale electrodeposition was first carried out by BP Solar [104] to achieve efficiencies above 10%. After 1984, a further research and development of CdTe electrodeposition was slowed down until recently, another breakthrough was found in electrodeposited CdTe in its post treatment by $MgCl_2$ by Major & coworkers [105] to achieve a new record of efficiency of 13%. Also, recently, Chauhan & co-worker have used ionic liquid based ED bath to deposit highly stoichiometric CdTe at lower process temperature [106].

4.2.2. Electrodeposition of CIS group of PV absorbers (CuInS₂, CuIn(S, Se)₂, Cu(InGa)Se₂)

The first ED attempt to produce a polycrystalline $CuInSe_2$ thin film of was reported by Bhattacharya [107] in which Cu, In, and Se were co-deposited from an acidic bath. Subsequently, EDA route with Cu-In alloys was attempted [108,109]. Bhattacharya *et al.* annealed electrodeposited In– Se/Cu – Se stacks [83]. The EDA routes which currently produce the most efficient $CuIn(S,Se)_2$ solar cells involve co-deposition of all of the constituent elements in complex medium [110] followed by annealing in a sulfur containing atmosphere, and also the ED of a Cu–In alloy with a small amount of Se incorporation from ion species, followed by annealing in sulfur. Both routes have produced CIS solar cells with about 11% efficiency. As band gap optimization plays a crucial role to improve efficiency, in the case of CIS system incorporating Ga has been found most successful in physical vapor deposited layer [111]. Calixto *et al.* first reported Ga incorporation in ED films [112] by co-deposition route.

Further reports on the electrodeposition of CIGS were limited due to the difficulties in controlling the deposition kinetics of the four species of wide-ranging potentials. The standard reduction potential values of Se^{4+}/Se, Cu^{2+}/Cu, In^{3+}/In and Ga^{3+}/Ga are +0.740, +0.34, -0.34 and -0.53 V vs. Saturated Hydrogen Electrode (SHE), respectively. The deposition potential of elemental Cu, In, Ga and Se, in case of electrodeposition of $Cu(In,Ga)Se_2$ can be obtained from the equilibrium potentials according to following reactions with accompanied Nernst equations:

Reaction	Corresponding deposition potential (vs. RHE)
$Cu^{2+}+2e^- \leftrightarrow Cu$	$V_{Cu^{2+}/Cu}=0.34+0.0295\log\left(a_{Cu^{2+}}/a_{Cu}\right)$
$In^{3+}+3e^- \leftrightarrow In$	$V_{In^{3+}/In}=-0.34+0.0197\log\left(a_{In^{3+}}/a_{In}\right)$
$Ga^{3+}+3e^- \leftrightarrow Ga$	$V_{Ga^{3+}/Ga}=-0.53+0.0197\log\left(a_{Ga^{3+}}/a_{Ga}\right)$
$HSeO_2^{\,+} + 4H^+ + 4e^- + OH^-$ $\leftrightarrow$ $HSeO_3 + 4H^+ + 4e^-$ $\leftrightarrow$ $Se + 3H_2O$	$V_{H_2SeO_3/Se}=V^0_{H_2SeO_3/Se}+$ $+\left(\frac{RT}{4F}\right)\log\left(a_{HSeO_2^+}/a_{Se}\right)+$ $+\left(\frac{3RT}{4F}\right)\log\left(C_{H^+}\right)=$ $=0.74+0.0148\log\left(a_{HSeO_2^+}/a_{Se}\right)-0.0433\ pH$

In above expressions, '*a*' refers to activity, for ions it is proportional to concentration of the respective ion and for pure elements it is considered to be unity.

This leads to a preferential deposition of single element causing overgrowing local non-stoichiometry. Due to the more negative reduction potential of Ga, its incorporation into the lattice becomes difficult. However, the best ED CIGS was reported by Kampmann *et al.* with a device efficiency of over 10% where they used a stacked layered scheme of ED [113]. The CIGS layer can be co-electrodeposited or it can be formed by a two stage approach. In co-deposition, the concentration and pH of the electrolyte needs to be adjusted or complex may be considered such that the electrode potentials of all the individual elements become closer to each other. The two stage strategy can be employed via stacked layer scheme with binary or ternary films, followed by an annealing in Se or S [114,115]. Co-deposition route is often preferred to reduce the harmful emissions, however, the prevailing factors in the co-deposition are largely: (i) the electrode potentials of individual ions in the electrolyte, (ii) cathodic polarization caused by the difference in deposition. potentials, (iii) relative ion concentrations in the electrolyte, (iv) the dissolution chances, and (v) the hydrogen overpotential on the deposited cathode surface. In most cases, the as-deposited films are found to be poorly crystalline and may consist of Cu–Se second phases, which may be reduced by a post-annealing treatment and brings about stoichiometry. Usually, the Cu/(In + Ga) and Ga/(In + Ga) ratios are critical in determining the cell performance [96]. Readers may refer to some of the well written reviews on electrodeposited CIS/CIGS solar cells for further details of synthesis and characterizations [117-121].

4.2.3. Electrodeposition of CZTS

An earth-abundant and environmentally benign Cu_2ZnSnS_4 (CZTS) are one of the ideal candidates for the production of thin film solar cells at large scale due to the large natural abundance of all the elements, a direct band gap in the range of 1.45 –1.6 eV [122,123], a high optical absorption coefficient of around 10^5 cm^{-1} and p-type semiconductor that matches well with the solar spectrum, and it has achieved a benchmark power conversion efficiency of more than 10% [124,125]. In EDA scheme, first the cathodic electrodeposition of constituent Cu, Zn and Sn can be expressed in following sets of reduction reactions and accompanied Nernst equations:

Reaction	Corresponding deposition potential (vs. RHE)
$Cu^{2+} + 2e \leftrightarrow Cu$	$E_{Cu^{2+} \to Cu} = 0.34 + 0.0295\ log[Cu^{2+}]$
$Zn^{2+} + 2e \leftrightarrow Zn$	$E_{Zn^{2+} \to Zn} = -0.76 + 0.0295\ log[Zn^{2+}]$
$Sn^{2+} + 2e \leftrightarrow Sn$	$E_{Sn^{2+} \to Sn} = -0.14 + 0.0295\ log[Sn^{2+}]$

Ennaoui *et al.* first reported a co-electrodeposition scheme for $Cu_2ZnSnSe_4$ alloy that was post-annealed using under H_2S environment at 550 °C for 2 h and the resulting solar cell showed an efficiency of 3.4% [126]. Addition of selenium to the system results in a band gap-lowering that helps to get a higher current density. Later, Scragg *et al.* attempted an EDA route with an electrodeposited metallic stack of Cu/Sn/Cu/Zn and its subsequent annealing in a sulfur-containing atmosphere mixed with 10% H_2/N_2 and heated to 575 °C for 2h to achieve a solar cell efficiency of 3.2% [127]. Redinger *et al.* in a novel approach annealed Cu and Zn precursor in a gas containing SnS and SnSe for 2 h at 540 °C to achieve efficiency 5.4% [128]. They pointed out that during annealing of a Cu/Zn precursor in the presence of SnS and SnSe vapor, the partial pressure of SnS and SnSe shifts the reaction equilibrium towards the formation of a single-phase $Cu_2ZnSn(S,Se)_4$ without Sn loss. However, a key drawback to some of the previous technique was longer duration of annealing, which is not desirable for commercial production of such CZTS absorbers. To address this issue, Zhang *et al.* [129] considered co-deposition of Cu-Zn-Sn in single bath followed by a rapid thermal annealing in Se atmosphere to obtain CZTSe solar cell with 4.5% power conversion efficiency. In a more deliberate study for CZTS system, Ahmed *et al.* [130] reported a shorter duration EDA route, where the metals stack of copper, zinc, and tin was sequentially electroplated on 600 nm sputtered molybdenum layer on soda lime glass substrate and annealed at low temperature (210–350 °C) in order to convert the elemental metals to metal alloys of CuSn and CuZn. After the formation of the homogeneous alloy of CuZn and CuSn layer, the film was given heat treatment at 585 °C for 12 min in a sealed quartz tube with 2–5 mg elementary sulfur under nitrogen atmosphere. In this study, they reported a champion Cu_2ZnSnS_4 solar cell with power conversion efficiency of 7.3%. The record efficiency is still below the bench mark at 11% obtained for CZTS solar cell by vacuum process [131]. One of the reasons could be the morphology and greater secondary phase fraction in EDA processed film. However, this efficiency can be improved further by considering other aspects of the solar cell device, such as reducing series resistance at the back contact, increasing shunt resistance by lowing deep level impurities, improving current density by reducing reflection and band-gap engineering and improving the open circuit voltage by thickness optimization and doping control.

4.2.4. Electrodeposition of other low cost PV

Other than the conventional non-silicon based earth abundant photovoltaic absorber, Zn_3P_2 is another promising compound semiconductor having a direct band gap of 1.5 eV [132]. Wyeth and Catalano [133] investigated the Schottky junction properties of Zn_3P_2 with various metals, and solar cells were constructed using metal/Zn_3P_2 Schottky contact by some researchers. The Mg/Zn_3P_2 solar cell shows the power conversion efficiency of 6% when the film was fabricated by gas phase growth, e.g. evaporation [134,135], MOCVD [136,137] and RF-sputtering [138]. Soliman *et al.* first reported an electrodeposited of Zn_3P_2 from aqueous solution containing zinc and phosphorus species [139] at normal temperatures and pressures. Later, Nose *et al.* reported electrodeposited Zn_3P_2 from aqueous solution where they constructed the Pourbaix (potential-pH) diagram of the Zn-P-H_2O system at 363 K and clarified the wide stable region of Zn_3P_2 at lower potentials [140]. They showed that, in order to obtain Zn_3P_2 by electrodeposition from aqueous solutions, it is necessary that the reduction of Zn^{2+} is suppressed and that of phosphate ions is enhanced.

5. Electrodepositon applied to dye sensitized solar cells

Dye Sensitized Solar Cell (DSSC)s are the classes of low cost photovoltaic device where a dye molecule (inorganic or organic) contributes to photoexcitation and transfer the photo-excited electron to the conduction band (CB) of a semiconductor having CB edge energetically favorable to accept the photo-excited electrons. Organic dye and pigment molecules are important class of materials for enhanced optical absorption in DSSCs due to their large oscillator strengths in aromatic systems. Therefore a lower film thickness can be adopted without compromising much on the efficiency. Two important approaches in this direction have been considered in DSSCs: (i) the use of multiple ultrathin films of high structural control using doping of layers to decrease the series resistance and (ii) the use of bulk heterojunctions consisting of a conductive pathway of a crystalline organic acceptor with large interfacial contact area to a matrix of a donor conductive polymer, prepared from a mixed solution of both constituents. An alternative way to make use of the strong absorption of organic dye molecules is realized in DSSCs [141,142] consisting of a porous wide-band gap semiconductor as electron-conducting phase, the organic dye as absorber, and a hole-conducting medium (typically iodine). The electron transport to the electrode in DSSCs is conventionally done by a TiO_2 layer. However, more economic solution at large scale could be to use ZnO as an alternative to TiO_2. The reason behind to consider ZnO is its versatility to form various forms of nanostructure and can be easily prepared by electrodeposition routes. The DSSCs can be realized based on porous ZnO as a wide-band gap semiconductor. As a post annealing step is not required, in general to obtain well-crystallized ZnO various types of substrates can be used. Additionally, ZnO as a thin film after Al-doping, Al:ZnO (AZO) serves as a transparent conductive oxide to replace more costly indium tin oxide (ITO) which is mainly due to the scarcity of In [89]. Other than the cost, another limitation of ITOs is inherent brittleness, that bars them to be grown on flexible substrates. In this aspect, the viability of ED of ZnO on flexible substrates can be appreciated. The combination of a compact AZO back electrode with electrodeposited porous ZnO layers for dye sensitization represents a very promising DSSC component with the prospect of low interface defects and production costs [143]. Different precursor routes have been attempted [144] to cathodically electrodeposit ZnO from aqueous electrolytes containing O_2, or NO^{3-}, or H_2O_2 as the source of oxygen. By reduction at the cathode, the [–OH] concentration rises in the solution. The reactions best describing this cathodic process can therefore be written as:

$$O_2 + 2H_2O + 4e \rightarrow 4OH^- \tag{13}$$

$$NO_3^- + H_2O + 2e \rightarrow NO_2^- + 2OH^- \tag{14}$$

$$H_2O_2 + 2e \rightarrow 2OH^- \tag{15}$$

Such an increase in pH leads to a situation when the solubility product of $Zn(OH)_2$ or ZnO is exceeded and a condition of supersaturation is achieved [69]. In the presence of appropriate crystallization seeds ZnO crystals nucleate and grow under controlled conditions. Readers may refer to the review by Yoshida and Schlettwein for further details of various possible hybrid inorganic-ZnO/ organic-dye structures applied to DSSCs [145], critical consideration of ZnO ED schemes by Yoshida et al. [146] and the detailed study of the role of structure directing agent by the same group with other workers [147]

6. Non-conventional electrodeposition: Electrodeposition from non-aqueous and ionic liquid bath

In the electrodeposition from aqueous solution, precursor films suffer from a number of problems. For example, in the case of $CuInS_2$ or $Cu(In,Ga)S_2$, the incorporation of In or Ga is not an easy task. Due to their limited solubility and large negative reduction potentials (very close to the hydrogen evolution reaction (HER) of water), it becomes necessary to use strong complexing agent to shift the reduction potential at more positive values. This may require toxic ligands, such as cyanide and to work under highly alkaline conditions to avoid HER which causes extremely poor morphology of the deposits. More unfortunately, In and Ga are often deposited together with their oxide or hydroxide forms under these conditions. A possible solution to these problems could be to replace the aqueous deposition medium by room temperature ionic liquids (RTIL). RTILs are defined as liquids consisting of molecular cations and anions that have a melting point below 100°C with ultra low vapor pressure. Due to the size of the molecular ions and the fact that the charge is delocalized through the molecule, RTILs have wide electrochemical windows. The ultra low vapor pressure and high degradation temperature allow them to be used at temperatures up to 250°C. The electrochemical window of most common pyrrolidinium-based RTILs are between 5 and 6 V [148]. Shivagan *et al.* [159] have shown that In and Ga can be reversibly electrodeposited from a eutectic based RTIL composed of a 1:2 choline chloride: urea mixture. El Abedin *et al.*studied the electrochemistry of Cu, In, and Se, individually in the 1-butyl-1- methylpyrrolidinium-bis(trifluoromethylsulfonyl)amide RTIL [150]. This group also showed that Ga can be deposited electrochemically on Au(111) electrodes [151] with this RTIL used. The high temperature ED using RTIL is interesting because a better crystallinity of the deposited films can be expected even without post heat treatment. Yang *et al.* [152] showed that for the electrodeposition of InSb from 1-ethyl-3-methylimidazolium chloride/tetrafluoroborate, at temperatures up to 120°C and the crystallinity of InSb film increased with the deposition temperature. The as-deposited InSb films were p-type and highly photoactive. First CdTe deposition onto glassy carbon, titanium, and tungsten substrates from 1-ethyl-3-methylimidazolium chloride/tetrafluoroborate in RTIL was carried out by Hsiu *et al* at 140°C [153]. Chauhan *et al.* [106] demonstrated that highly stoichiometric *p*-type CdTe absorber can be grown using BMIM Cl⁻ RTIL bath. The electrodeposited *p*-CZTS absorber layer from RTIL was first attempted by Chan *et al.* using a eutectic mixture of choline chloride and ethylene glycol at a molar ratio of 1:2 [154] where Cu, Sn and Zn could be deposited at potentials -0.55V, -0.67V and -1.1V (vs. Ag/AgCl) in a cathodic scan. The resulting CZTS film after annealing was highly crystalline and free from the incorporation of oxide impurities that usually appears in the case of aqueous medium.

7. Conclusions

Electrodeposition, a low cost non-vacuum technique for fabrication of thin films deserves its place in the development of next generation solar cells and modules. Owing to its simplicity and scalability, various compound semiconductor photovoltaic absorbers, such as Cu_2ZnSnS_4, Cu_2S, SnS, Cu_2SnS_3 *etc.* have already been fabricated at lab and pilot scales. Although, these photovoltaic absorbers may suffer poor phase purity and crystalline nature, controlling the deposition parameters and surface functionalization of the deposited layer can help achieve ultra low cost and environment friendly thin film solar cells and modules.

Author details

Abhijit Ray*

Address all correspondence to: abhijit.ray1974@gmail.com

School of Solar Energy, Pandit Deendayal Petroleum University, Raisan, Gandhinagar, Gujarat, India

References

[1] Fthenakis V. (2009). Ren. Sust. Energy Rev. 13(9), 2746–2750.

[2] Alonso, E., et al. (2007). Env. Sci. & Technol. 41(19), 6649–6656.

[3] Ginley, D., et al. (2008). MRS Bull. 33, 355.

[4] Habas, S.E., et al. (2010). Chem. Rev. 110, 6571–6594.

[5] Würfel, P. and Würfel, U. (2009). Physics of solar cells: from basic principles to advanced concepts. John Wiley & Sons.

[6] T. Unold and H.W. Schock. (2011). Annu. Rev. Mater. Res. 41, 297.

[7] Chopra, K. L. and Das, S.R. (1983). Why Thin Film Solar Cells?. Springer, USA.

[8] López, N., et al. (2011). Phys. Rev. Lett. 106.2, 028701.

[9] Shafarman, W. N., Klenk, R. and McCandless, B.E. (1996). J. Appl. Phys. 79.9, 7324-7328.

[10] Romeo, N., et al. (1999). Solar Energy Mater. Sol. Cells 58.2, 209-218.

[11] Guillemoles, J-F., et al. (2000). J. Phys. Chem. B, 104.20 (2000): 4849-4862.

[12] Fujii, M., et al. (1988). Solar Energy Mater., 18, 23.

[13] Romeo, N., et al. (1978) Appl. Phys. Lett., 32, 807.

[14] Jones, A. C. (1997). Chem. Soc. ReV., 26, 101.

[15] O'Brien et al. (1996). J. Cryst. Growth, 167, 133.

[16] Fainer et al. (1996). Thin Solid Films, 280, 16.

[17] Cheon et al. (1997). J. Am. Chem. Soc., 119, 3838.

[18] Fujita, S., et al. (1996). Cryst. Growth, 164, 196.

[19] Fraser, D. B. and Melchior, H. J. (1972). Appl. Phys., 43, 3120.

[20] Liu et al. (1994). J. Appl. Phys., 75, 3098.

[21] Meng et al. (1995). Vacuum 1995, 46, 1001

[22] Asom, M. T., et al. (1991). J. Crystal Growth 112.2, 597-599.

[23] Fraas, L. M., et al. (1986). J. Electron. Mater. 15.3, 175-180.

[24] Sivakov, V., et al. (2009). Nano Lett. 9.4, 1549-1554.

[25] Gur, I., et al. (2005). Science 310.5747, 462-465.

[26] Kyoohee, W., et al. (2012). Energy & Env. Sci. 5.1, 5340-5345.

[27] Maeda K, et al. (2011) Solar Energy Mater. Sol. Cells 95: 2855–2860.

[28] Akhavan,V.A., et al. (2012). J. Solid State Chem. 189, 2–12.

[29] Woo K, Kim Y, Moon J. (2012). Energy Environ. Sci. 5: 5340-5345

[30] Rajeshmon, V.G., et al. (2011). Solar Energy 85, 249–255.

[31] Das, S., et al. (2012). ECS Trans. 45 (7), 153-161.

[32] Patel, M., et al. (2012). J. Phys. D: Appl. Phys. 45, 445103.

[33] Mali. S.S., et al. (2012). Electrochim. Acta. 66, 216– 221.

[34] Wangperawong, A., et al. (2011). Thin Solid Films 519, 2488–2492.

[35] Scragg, J.J., et al. (2010). J. Electroan. Chem. 646, 52–59.

[36] Ennaoui, A., et al. (2009). Thin Solid Films 517: 2511–2514.

[37] Lincot, D. (2005). *Thin Solid Films* 487.1, 40-48.

[38] Fulop, G. F. and Taylor, R.M. (1985). *Annual Rev. Mater. Sci.* 15.1, 197-210.

[39] Gary, H., ed. (2008). Electrochemistry of Nanomaterials. John Wiley & Sons.

[40] Ruythooren, W., et al. (2000). *J. Micromech. Microengg.* 10.2, 101.

[41] Dharmadasa, I. M., and Haigh, J. (2006). *Journal of The Electrochemical Society* 153.1, G47-G52.

[42] Scragg, J.J., et al. (2008). Electrochem. Commun. 10.4, 639-642

[43] Paunovic, M. and Schlesinger, M. (2006). FUNDAMENTALS OF ELECTROCHEMI-CAL DEPOSITION, Wiley Interscience, 2nd Ed., New Jersey.

[44] J. Choi, R. B. et al. (2004). Electrochim. Acta, 49, 2645.

[45] Y. T. Sul, et al. (2001). Medical Engg. Phys., 23, 329.

[46] Feng, X., et al. (2008). Nano Letters, 8.11, 3781-3786.

[47] Baxter, J. B., et al. (2006). Nanotechnology, 17.11, S304.

[48] Mor, G. K., et al. (2005). Adv. Func. Mater. 15.8, 1291-1296.

[49] Yu, X., et al. (2008). Sensors and Actuators B: Chemical, 130.1, 25-31.

[50] Chu, S-Z., et al. (2005). Adv. Func. Mater. 15.8, 1343-1349.

[51] Mor, G. K., et al. (2006). Nano Letters 6.2, 215-218.

[52] Paulose, M., et al. (2006). J. Phys. D: Appl. Phys. 39.12, 2498.

[53] Leenheer, A. J., et al. (2007). J. Mater. Res. 22.03, 681-687.

[54] Macak, J. M., et al. (2006). Chem. Phys. Lett. 428.4, 421-425.

[55] Premchand, Y. D., et al. (2006). Electrochem. Commun. 8.12, 1840-1844.

[56] Yu, X., et al. (2006) Nanotechnology 17.3, 808.

[57] Huang, H., et al. (2013). Energy Environ. Sci. 6.10, 2965-2971.

[58] Wang, C. E., et al. (2014). J. Mater. Sci. Nanotech. 1.1, 1.

[59] Owen, J. I., et al. (2011). Phys. status solidi (a) 208.1, 109-113.

[60] Dennison, S., (1993). Electrochim. Acta 38, 2395.

[61] H. Gobrecht, H.D., et al. (1973). Phys. Chem. 67, 930.

[62] Pourbaix, M., (1974). Atlas of electrochemical equilibria in aqueous solutions. National association of corrosion engineers (2nd English Edn.) USA.

[63] Savadogo, O., (1998). Solar Energy Mater Sol. Cells 52, 361–388.

[64] Dremlyuzhenko, S.G., et al. (2008). Inorg Mater 44, 21–29.

[65] Bouroushian, M., (2010). Electrochemistry of Metal Chalcogenides, Springer-Verlag Berlin Heidelberg.

[66] Tsao, J. Y., (1993). Materials Fundamentals of Molecular Beam Epitaxy, Academic Press, London.

[67] Stringfellow, G. B., (1989). Organometallic Vapor-Phase Epitaxy: Theory and Practice Academic Press, Inc., London.

[68] Sze, S. M. (1985). Semiconductor Devices Physics and Technology.

[69] Gregory, B.W. and Stickney, J.L. (1991). J. Electroanal. Chem. 300, 543.

[70] Innocenti, M., et al. (2001). J. Electrochem. Soc. 148.5, C357-C362.

[71] Pezzatini, G., et al. (1999). J. Electroanal. Chem., 475, 164.

[72] Gregory, B. W., et al. (1990). J. Electroanal. Chem., 293, 85.

[73] Gregory, B. W. and Stickney, J. L. (1991). J. Electroanal. Chem., 300, 543.

[74] Demir, U. and Shannon, C. (1994). Langmuir, 10, 2794.

[75] Streltsov, E. S., et al. (1994). Dokl. Akad. Nauk Bel., 38, 64.

[76] Foresti, M. L. et al. (1998). J. Phys. Chem. B 102, 38, 7413.

[77] Wade, T. L., et al. (1999). Electrochem. Solid-State Lett., 2, 616.

[78] Innocenti, M., et al. (2001). Electroanal. Chem., 514, 75.

[79] Colletti, L. P. and Stickney, J. L. (1995). Abstracts of the 210th ACS National Meeting, Chicago, IL, August 20-24, 1995, (Pt. 1), INOR-297.

[80] Technology Roadmap, Solar photovoltaic energy, IEA Publication, 2014.

[81] Holden, N. E., "Table of the Isotopes", in Lide, D. R., Ed., CRC Handbook of Chemistry and Physics, 86th Ed., CRC Press, Boca Raton FL, 2005.

[82] Alharbi, F. et al. (2011). Renewable Energy 36, 2753e2758.

[83] Bhattacharya, R.N., et al. (1996) J. Electrochem. Soc., 143, 854.

[84] Kroger, F.A. (1978). J. Electrochem. Soc., 125, 2028.

[85] Voss, T., et al. (2007). 22nd European Photovoltaic Solar Energy Conference, 3 – 7 September 2007, Milan.

[86] Zank, J., et al. (1996). Thin Solid Films, 286, 259.

[87] Kampmann, A., et al. (2003). Proceeding of the MRS Spring Meeting, San Francisco.

[88] Scragg, J.J., et al. (2010). J. Electroanal. Chem., 646, 52.

[89] Scragg, J.J., et al. (2008). Electrochem. Commun., 10, 639.

[90] Scragg, J.J., et al. (2009). Thin Solid Films, 517, 2481.

[91] Scragg, J.J., et al. (2008). Phys. Stat. Sol. (b), 245, 1772.

[92] Araki, H., et al. (2009). Solar Energy Mater. Solar Cells, 93, 996.

[93] Guillen, C. and Herrero, J. (1996). J. Electrochem. Soc., 143, 493 – 8.

[94] Panicker, M.P.R., et al. (1978). J. Electrochem. Soc., 125, 566.

[95] Basol, B.M. and Tseng, E.S.F. (1983). J. Electrochem. Soc., 130, C243.

[96] Dale, P.J., et al. (2008). J. Phys. D, 41, 085105.

[97] Lincot, D. et al (2004). Solar Energy,77, 725.

[98] Calixto, M.E., et al. (2005). Conference Record of the 31st IEEE Photovoltaic Specialists Conference, p. 378.

[99] Basol, B.M., et al. (1984). J. Appl. Phys., 58, 3809.

[100] Rajeshwar, K., and Bhattacharya, R.N. (1984). J. Electrochem. Soc., 131, C314.

[101] Rajeshwar, K., et al. (1984). J. Electrochem. Soc., 131, C313.

[102] Bhattacharya, R.N. and Rajeshwar, K. (1985). J. Appl. Phys., 58, 3590.

[103] Bhattacharya, R.N., et al. (1985). J. Electrochem. Soc., 132, 732.

[104] Turner, A.K. et al, (1994). Solar Energy Mater. Sol. Cells, 35, 263.

[105] Major, J. D., et al. (2014). Nature 511, 334-337.

[106] Chauhan, K.R. et al. (2014). J. Electroanal. Chem. 713, 70–76.

[107] Bhattacharya, R.N. (1983) J. Electrochem. Soc., 130, 2040.

[108] Hodes, G., et al. (1985). Thin Solid Films, 128, 93.

[109] Kapur, V.K., et al. (1987) Solar Cells, 21, 65 – 72.

[110] Kessler et al. (2005) 20th European Photovoltaic Solar Energy Conference, 6 – 10 June 2005, Barcelona.

[111] Bar, M., et al. (2009) Appl. Phys. Lett., 95, 3.

[112] Calixto, M.E., et al. (1999) Solar Energy Mater. Sol. Cells, 59, 75.

[113] Kampmann, A., et al. (2003) Proceeding of the MRS Spring Meeting, San Francisco.

[114] Zank, J., et al. (1996). Thin Solid Films 286, 259–263.

[115] Friedfeld, R., et al. (1999). Solar Energy Mater. Sol. Cells 58, 375–385.

[116] Bhattacharya, et al. (1998) Solar Energy Mater. Sol. Cells 55, 83–94.

[117] Lincot, D., et al. (2004). Sol. Energy 77, 725–737.

[118] Kaelin, M., et al. (2004) Sol. Energy 77, 749–756.

[119] Lincot, D., (2005). Thin Solid Films 487, 40–48.

[120] Hibberd, C.J., et al. (2010). Prog. Photovolt. Res. Appl. 18, 434–452.

[121] Saji, V.S., et al. (2011). Solar Energy 85, 2666–2678.

[122] Jimbo, K. et al. (2007). Thin Solid Films, 515, 5997.

[123] Ito, K. and Nakazawa, T. (1988). Jpn. J. Appl. Phys. 27, 2094.

[124] Mitzi, D.B. et al. (2011). Sol. Energy Mater. Sol. Cells, 95, 1421.

[125] Todorov, T.K. et al. (2010). Adv. Mater. 22, 1.

[126] Ennaoui, A., et al. (2009). Thin Solid Fims, 517, 2511–2514.

[127] Scragg, J. J., et al. (2010). J. Electroanal. Chem., 646, 52.

[128] Redinger, A., et al. (2011). J. Am. Chem. Soc. 2011, 133, 332.

[129] Zhang, Y.Z., et al. (2013) Solar Energy, 94, 1-7.

[130] Ahmed, S., et al. (2012). Adv. Energy Mater., 2, 253–259.

[131] Torodov, T.K. et al. (2013). Adv. Energy Mater. 3, 34–38.

[132] Fagen, E. A., (1979). J. Appl. Phys., 50, 6505

[133] Wyeth, N. C. and Catalano, A. (1980). J. Appl. Phys., 51, 2286

[134] Lousa, A. et al. (1985). Sol. Energy Mater., 12, 51

[135] Murali, K. R., (1987) Mater. Sci. Eng., 92, 193.

[136] Long, J. (1983). J. Electrochem. Soc., 130, 725.

[137] Hermann, A. M., et al. (2004). Sol. Energy Mater. Sol. Cells, 82, 241

[138] Suda, T., et al. (1988). J. Cryst. Growth, 86, 423

[139] Soliman, M., et al. (2005). Renewable Energy, 30, 1819

[140] Nose, Y., et al., (2012). Journal of The Electrochemical Society, 159 (4) D181-D186

[141] Hagfeldt, A. and Grätzel, M. (2000) Molecular photovoltaics. Acc. Chem. Res., 33 (5), 269 – 277.

[142] Grätzel, M. (2001). Nature, 414, 338 – 344.

[143] Loewenstein,T., et al. (2008). Phys. Stat. Sol. (a), 205, 2382 – 2387.

[144] Advances in Electrochemical Science and Engineering, Edited by Richard C. Alkire, Dieter M. Kolb, Jacek Lipkowski, and Philip N. Ross, Wiley (2010).

[145] Yoshida, T. and Schlettwein, D. (2004) Electrochemical self - assembly of oxide/ dye composites, in Encyclopedia of Nanoscience and Nanotechnology, vol. 2 (ed. A. Nalwa), American Scientifi c Publishers, Stevenson Ranch, pp. 819 – 836.

[146] Yoshida, T., et al., (1999). Chem. Mater., 11, 2657 – 2667.

[147] Yoshida, T., et al., (2002). Electrochemistry, 70, 470 – 487

[148] Abedin, S.Z.E. and Endres, F. (2006) Chem.Phys.Chem., 7, 58.

[149] Shivagan, D.D., et al. (2007). Thin Solid Films, 515, 5899.

[150] Abedin, S.Z.E., et al. (2007) Electrochim. Acta, 52, 2746.

[151] Gasparotto, LHS., et al. (2009) Electrochim. Acta, 55, 218.

[152] Yang, M.H., et al. (2003) J. Electrochem. Soc., 150, C544.

[153] Hsiu, S.I. and Sun, I.W. (2004) J. Appl. Electrochem., 34, 1057.

[154]   Chan, C.P., et al., (2010). Solar Energy Mater. Sol. Cell 94, 207

[155]   Anuar, K., et al. (2002). Solar Energy Mater. Sol. Cell. 73, 351.

[156]   Koike, J., et al. (2012). Jap. J. Appl. Phys. 51.10S, 10NC34.

[157]   Mathews, N. R., et al. (2013) J. Mater. Sci.: Mater. Electron. 24.10, 4060-4067.

[158]   Yong-Sheng, C., et al. (2012) Chinese Phys. B 21.5, 058801.

[159]   Cui, Y., et al. (2011) Solar Energy Mater. Sol. Cells 95.8, 2136-2140.

[160]   Pawar, B. S., et al. (2011). ISRN Renewable Energy 2011.

[161]   Jeon, J-O., et al. (2014). ChemSusChem 7.4, 1073-1077.

[162]   Guo, L., et al. (2014). Prog. Photovoltaics: Res. Appli. 22.1 (2014): 58-68.

[163]   Septina, W., et al. (2013). Electrochimica Acta 88, 436-442.

[164]   Li, J., et al. (2012) Appl. Surf. Sci. 258.17, 6261-6265.

[165]   Steichen, M. et al. (2013). J. Phys. Chem. C, 117 (9), 4383–4393.

[166]   Ghazali, A., et al. (1998). Solar Energy Mater. Sol. Cells, 55.3, 237-249.

[167]   Subramanian, B., et al. (2001). Mater. Chem. Phys., 71.1, 40-46.

[168]   Zainal, Z., (1996). Solar Energy Mater. Sol. Cells, 40.4 (1996): 347-357.

[169]   Cheng, S., et al. (2006) Thin Solid Films 500.1, 96-100.

[170]   Cheng, Shuying, et al. (2007). Mater. Lett. 61.6, 1408-1412.

[171]   Cheng, Shuying, et al. (2006). Opt. Mater. 29.4, 439-444.

[172]   Mkawi, E. M., et al. (2014). Solar Energy Mater. Sol. Cells 130, 91-98.

6

Pulse Electroplating of Ultrafine Grained Tin Coating

Ashutosh Sharma, Siddhartha Das and Karabi Das

Abstract

In the electronic packaging industries, soldering materials are essential in joining various microelectronic networks. Solders assure the reliability of joints and protect the microelectronic packaging devices. They provide electrical, thermal, and mechanical continuity among various interconnections in an electronic device. The service performance of all the electronic appliances depends on high strength and durable soldering materials. Lead-containing solders are in use for years, resulting in an extensive database for the reliability of these materials. However, due to toxicity and legislations, lead-free solders are now being developed. As tin (Sn) is the major component of solders, this chapter presents the detailed results and discussion about the metallurgical overview of Sn, synthesis, and characterization of pulse electrodeposited pure tin finish from different aqueous solution baths. The experiments on pulse electrodeposition such as common tin plating baths employed, their chemical compositions, rationale behind their selection and their characterization by bath conductivity and cathodic current efficiency, microstructures, and tin whisker growth are discussed. Further, the effect of pulse electrodeposition parameters such as current density, additive concentration, pH, duty cycle, frequency, temperature, and stirring speed on microstructural characteristics of the coating obtained from sulfate bath and their effect on grain size distribution have been presented.

Keywords: Tin plating, pulse electrodeposition, morphology, plating baths, grain size

1. Introduction

1.1. Background

Lead-bearing solders are in use till date due to their indispensable properties [1–3]. Firstly, it is very cheap, available in abundance and provides good physical and chemical bonding to the substrate without interfering with the substrate [4–5]. Secondly, lead (Pb) reduces the surface tension of pure tin, which is 550 mN/m at 232°C, and the lower surface tension of 63Sn-37Pb solder (470 mN/m at 280°C) facilitates wetting [4, 6, 7]. The presence of Pb also helps to prevent the white tin (β-Sn) to gray tin (α-Sn) allotropic phase transformation from occurring in high-Sn solders upon cooling below 13ºC [7]. This makes it a prime choice for soldering.

1.2. Replacement of Pb

There are many technology-based problems that can serve as reasons for the elimination of Sn-Pb solders. First, it has been already proved in the past that many lead-free candidate solders exhibit significantly better strength and fatigue life properties [4, 6]. Secondly, Pb and Pb compounds have been cited by the Environmental Protection Agency (EPA) as one of the top 17 chemicals posing the greatest threat to human life and the environment [7, 8]. In view of these reasons, elimination of Pb from electronics is necessary in electronics packaging [1, 6].

1.3. The lead-free definition and regulations

Legislations to restrict the use of Pb were first implemented in the USA in 1991 with the Lead Exposure Reduction Act of 1991 and the Lead Exposure Act of 1992, which bans Pb in some applications and limits Pb content in others to less than 0.1% [7, 8]. In the United States, the National Electronics Manufacturing Initiative (NEMI) program was developed to research on lead-free alternatives [9]. In Japan, this movement is connected to the Lead-free Soldering Research Council (1994 to 2000) within the Japan Institute of Electronic Packaging. Japan Electronics and Information Technology Industries Association (JEITA) has set guidelines for lead-free products, which was published in 1999 [10, 11]. These companies aimed to use lead-free solders in mass-produced consumer products and to implement lead-free soldering technologies in their products by 2003 [2]. In Europe, the EU directives (WEEE) and (RoHS) have issued a ban on the use of lead in consumer goods [12, 13].

2. Metallurgical overview of Sn

Pure Sn has two allotropes, white tin (β-Sn, metallic) and gray tin (α-Sn, semiconductor). The most common form of Sn in real life is white tin, which transforms to gray tin at temperatures below 13ºC [14]. β-Sn possesses body-centered tetragonal crystal structure with lattice parameters $a = b = 0.5820$ nm and $c = 0.3175$ nm [15]. This c/a ratio of 0.546 gives rise to highly anisotropic behavior in Sn. The mechanical properties are very poor, the hardness being 11 Hv and tensile strength 44 MPa [16]. This is largely due to its lower melting point relative to common engineering metals such as aluminum (Al), copper (Cu), and steel [14–16].

2.1. Pb-free Sn alloys

A relatively large number of Pb-free solder alloys have thus been proposed so far, Sn being the primary or major constituent. The two other elements that are major constituents are indium (In) and bismuth (Bi). Other alloying elements are zinc (Zn), silver (Ag), antimony (Sb), copper (Cu), and magnesium (Mg), and in one case, a minor amount of Pb [1–3]. The most popular Pb-free alloy system candidates are listed in a thorough review paper by Abtew and Selvaduray [1]. The most important characteristics that must be considered in selecting suitable Pb-free solder candidates are: nontoxic; availability; sufficient electrical as well as thermal conductivity; adequate mechanical properties compatible with metallic substrates such as Cu, nickel (Ni), Ag, or gold (Au); economically viable; acceptable melting and processing temperatures; and less temperature effects on substrates, printed circuit boards(PCBs), etc. [1–4, 6].

2.2. Intermetallic compounds and whisker growth

Due to the Pb-free solder implementation, pure Sn and Sn-Cu, Sn-Ag are commonly used to replace eutectic Sn-Pb as the surface finish on the lead-frames and metal terminations of passive devices [17, 18, 19]. However, in Sn-rich lead-free finish, Sn whiskers have been found to form, which poses a serious threat for the reliability of passive devices. The Sn whisker formation was first reported in 1946 [20]. It is generally accepted that the driving forces of Sn whiskers mainly attribute to the internal stresses, the dissolution of the metal under-layer, and the interfacial compound formation. There are two intermetallic compounds of Cu and Sn below 300°C, they are Cu_6Sn_5 and Cu_3Sn. The Cu_6Sn_5 forms at room temperature while Cu_3Sn forms after annealing at elevated temperatures [21, 22]. Some intermetallic compounds (IMCs), such as AuSn4 for Sn/Au couples, Cu_6Sn_5 for Sn/Cu couples, and Ni_3Sn_4 for Sn/Ni couples, even form at room temperature [23, 24].

Based on these three main root-causes of Sn whisker formation, three methods are devised by researchers to retard the Sn whiskers formation: (1) choosing optimal thickness of the finish layer, (2) alloying with other metal elements, and (3) adding a reaction barrier layer beneath the finish layer [25–27].

3. Synthesis routes and technologies for solder fabrication

There are many synthesis routes for the fabrication of solders varying from solid state mechanical alloying, powder metallurgy, sol gel, melting and casting route, chemical routes, gaseous phase sputtering or evaporation methods, electrodeposition method, etc. [28-35]. Among all these methods, we will discuss techniques related to thin film pulse electrodeposition of solders. The other methods are not discussed here because they are out of the scope of this chapter.

3.1. Electrodeposition

Both evaporation and sputter deposition techniques require high vacuum and/or high temperature processing, which increases operation costs and cause inter-diffusion problems. Compared to these fabrication processes, electrodeposition is an economically viable process [36]. It can be used to plate either single layer or multilayer deposits with easy and precise control of the thickness and composition of each layer. Electrodeposition can be performed on substrates with varying sizes and complex shapes and the deposit can be very thin or very thick [37–38]. There are, however, safety and environmental concerns related with chemical treatment and safe disposal of wastes. Numerous metals and metal alloys have been successfully electrodeposited from aqueous solutions. The most useful electrodeposited metals include Sn, Cr, Cu, Ni, Ag, Au, Zn, and alloys such as chromium-nickel (Cr-Ni), iron-cobalt (Fe-Co), and various Sn alloys [36–38].

3.2. General aspects of electrodeposition

In electrodeposition, metal ions present in a solution, the electrolyte, are reduced at the surface of an electrode to form a metal layer, as shown in Fig. 1.

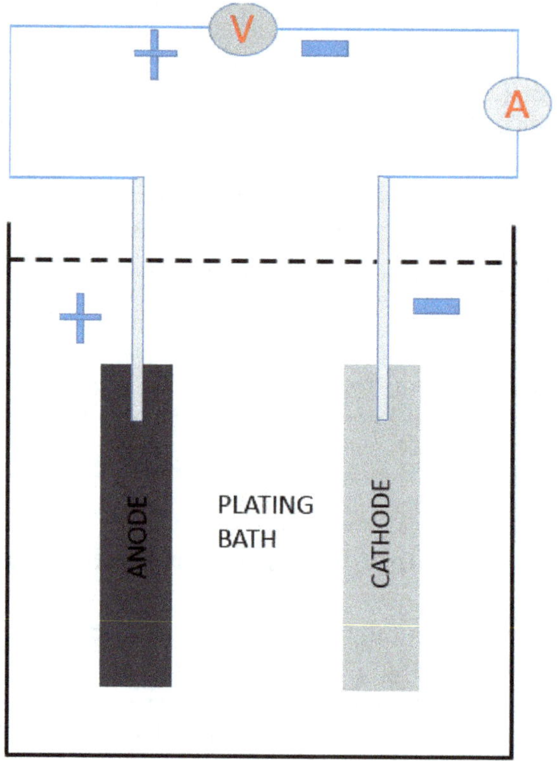

Figure 1. A schematic for the electrodeposition process.

This process essentially consists of: (1) an anode (the positive electrode), (2) a cathode (the item to be electroplated, which is the negative electrode), (3) the electrolyte acts as a transport medium for the tin ions to be deposited at the cathode as a coating on the item to be electroplated, (4) an electric current or voltage source for controlling the deposition, and (5) various peripherals for contacting the electrodes, stirring and heating the solution, etc. [36–37].

In electrodeposition, the metal is deposited over a conductive substrate by the application of electric current through the electrolytic bath. The current provides sufficient energy to proceed the oxidation-reduction reactions at anode and cathode, respectively. The metal ions in the electrolyte accept the electrons and get deposited on the substrate. The weight of the deposited material can be calculated from the relation given by Faraday's laws [37–39]:

$$\frac{\text{Thickness}}{\text{time}} = \frac{J(ECE)(CCE)}{\text{density}} \tag{1}$$

where thickness of the deposit is in mg, time in seconds, J = current density, ECE = electrochemical equivalent, CCE = current efficiency (ratio of actual/theoretical weight deposited), and density of deposit is in g/cm^3.

3.3. Cathodic and anodic reactions

Electrochemical deposition of metals and alloys involves the reduction of metal ions from aqueous, organic, and fused-salt electrolytes. In this thesis, electrodeposition from aqueous solutions is being considered. The reduction of metal ions M^{n+} in aqueous solution is represented by [37–39]:

$$M^{n+} + ne^- \rightarrow M \tag{2}$$

Reaction (2) is often accompanied by hydrogen evolution. In acid solutions, we have:

$$2H^+ + 2e^- \rightarrow H_2 \tag{3}$$

In neutral and basic solutions, hydrogen evolution follows the equation:

$$2H_2O + 2e^- \rightarrow H_2 + 2OH^- \tag{4}$$

At the anode, the anodic reactions are as follows. In an acidic solution, we have:

$$2H_2O \rightarrow O_2 + 4H^+ + 4e^- \tag{5}$$

In an alkaline solution, the anode reaction is:

$$4OH^- \rightarrow O_2 + 2H_2O + 4e^- \tag{6}$$

For a soluble anode, oxidation reactions, which will dissolve the anode into solution:

$$M \rightarrow M^{n+} + ne^- \tag{7}$$

The equilibrium electrode potential between a metal and a solution of its ions is given by the Nernst equation:

$$E_{M/M^{n+}} = E^o + \frac{RT}{nF} \ln C_{M^{n+}} \tag{8}$$

where E^o is the standard electrode potential, R is the gas constant, F is the Faradaic constant, and $C_{M^{n+}}$ is the metal ion concentration. It can be noticed that the value of $E_{M/M^{n+}}$ depends on the concentration of metal ions to be plated, addition of additives, current density, etc. [38, 39].

3.4. Electrolyte conductivity

The conductivity of an electrolyte depends on the degree of dissociation, the mobility of individual ions, temperature (and thus viscosity), and the electrolyte composition. In aqueous solutions, the ionic conduction depends on the degree of dissociation of dissolved species in the solution [38, 39]. In order to increase the conductivity of electrolytes, certain salts and acids or alkali are added; these are known as supporting electrolytes. For acid electrolytes, chlorides and acids are useful; for neutral electrolytes, chlorides are useful; and for alkaline electrolytes, sodium hydroxide or cyanides are useful [39]. Generally, a marked decrease in conductivity at higher concentration is due to the greater coulombic forces acting between the ever closer ions in solution. This leads to the loose association of opposite charged ions that are effectively neutral and thus no longer contribute to the overall conductivity [39].

3.5. Polarization and overpotential

The equilibrium potential of an electrode differs in an electrochemical cell after the application of current. Suppose the equilibrium potential of an electrode when there is no external current flowing is E. After the application of an external current (I), the potential of the electrode change by E(I), the overpotential (η) can be given as:

$$\eta = E(I) - E \tag{9}$$

It can be also expressed as in terms of current and voltage by the Tafel equation:

$$\eta = a + b \log I \tag{10}$$

where a and b are constants.

The overpotential η is required to overcome hindrance of the overall electrode reaction, which is usually composed of a sequence of partial reactions, charge transfer, diffusion, chemical reaction, and crystallization [36, 38]. The overpotential is generally used as a measure of the extent of polarization. Polarization demonstrates departure of the electrode potential from the equilibrium value upon passage of a Faradaic current [39]. Depending upon the nature and sites of the inhibition factors, there may be different kinds of overpotentials, such as charge transfer overpotential or activation overpotential, which is caused by inhibition of the charge transfer taking place across the close proximity of solution/cathode interface.

Thus, four different kinds of overpotential are distinguished and the total overpotential η can be considered to be composed of four components: where η_{ct}, η_d, η_r, and η_c are, as defined above, charge-transfer, diffusion, reaction, and crystallization overpotentials, respectively. The electrode reaction can then be represented by a resistance (R_E), composed of a series of resistances (or more precisely, impedance), which represent the different steps: R_{mt}, R_{ct}, and R_{rxn}. A fast reaction step is characterized by a small resistance, while a slow reaction step is represented by a high resistance. In other words, R_E will be the sum of all the resistances R_{mt}, R_{ct}, and R_{rxn}, as shown in Fig. 2 [38].

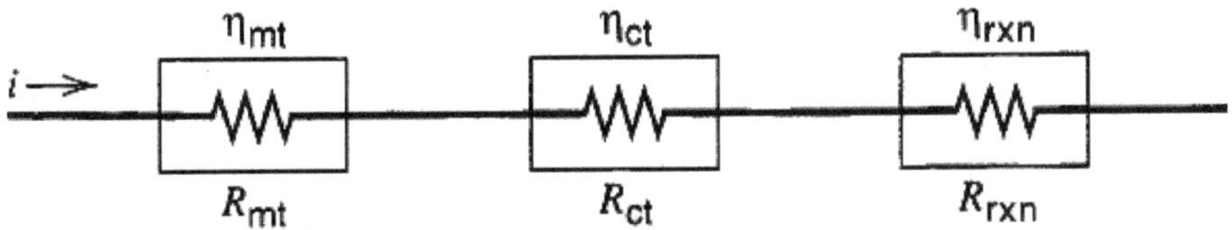

Figure 2. Processes in an electrode reaction represented as resistances (From A.J. Bard, and L.R. Faulkner, Chapter 1: Introduction and Overview of Electrode Processes, Page 24, Electrochemical Methods - Fundamentals and Applications, 2nd Edition.

3.6. Pulse current electrodeposition

In olden times, direct current (DC) electrodeposition had only one parameter, namely, current density that is variable. In modern times, pulsed current (PC) plating where the potential or current density alternates rapidly between two different values is used [36–39]. This is accomplished with a series of pulses of equal amplitude, duration, and polarity, separated by a period of zero current, and time (t) axis as shown in Fig. 3. Each pulse consists of an on-time (T_{on}), during which potential and current is applied, and an off-time (T_{off}), during which open circuit potential and zero current is applied.

The duty cycle is given by the equation:

$$\text{Duty Cycle }(\%) = \frac{T_{on}}{T_{on} + T_{off}} \tag{11}$$

The average current density is defined as:

$$J_{average} = (J_{peak}) \times \text{Duty Cycle} \tag{12}$$

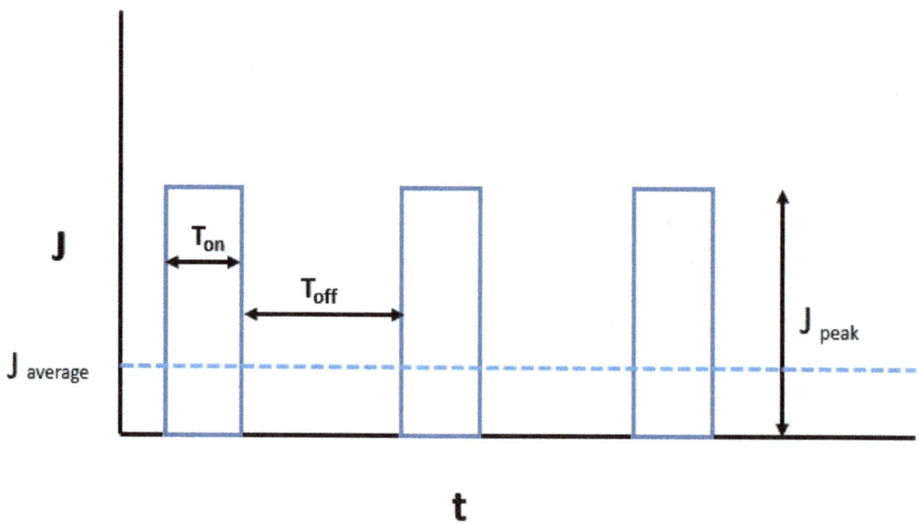

Figure 3. Schematic diagram of pulsed current waveform.

$J_{average}$ is the average current density and J_{peak} is the peak current density [35, 39, 40]. The average current density in PC plating is similar to the current density used in DC plating. For a given average current density, a number of combinations of different peak current densities and on-off times are available. This gives PC plating two important features. Firstly, a very high instantaneous current density, one to two orders in magnitude greater than the steady-state limiting current density as reported, can be used without depleting the metal ions at the electrode surface [40]. This favors the initiation of the nucleation process and greatly increases the number of grains per unit area, resulting in a finer grained deposit with better properties than conventional DC plating. Secondly, during the off-time period, adsorption and desorption, as well as recrystallization of the deposit occurs, which can control the microstructure of the deposits [36, 40, 41].

4. Electrodeposition of Sn

Sn can be electrodeposited from both acidic and alkaline aqueous solutions [42, 43]. In acidic baths, Sn usually exists as Sn^{2+} ions, while in alkaline baths, Sn^{4+} is the more stable species. The various Sn plating baths available in literature are discussed below.

4.1. Sodium stannate baths

The alkaline stannate electrolytes usually contain sodium or potassium stannate and the corresponding alkali metal hydroxide. These electrolytes are environment friendly, as they are non-corrosive in nature and do not require other organic additives [42, 43]. The alkaline tin baths have a very high throwing power and can be operated over a wide current density range.

The tin coatings electroplated from alkaline electrolytes possess improved solderability, since they do not require any organic additive agents [44, 45]. This leads to a great improvement in wettability. One major disadvantage of alkaline plating baths is that highly alkaline baths may dissolve the photoresist used to define areas on semiconductor wafers. Alkaline plating baths also usually require higher plating temperatures (60–70°C) as compared to acidic ones [46].

4.2. Stannous sulfate baths

Sulfate baths are primarily used for bright acid Sn plating. Organic agents are necessary if bright and dense films are to be obtained. Electroplating of tin from acidic stannous (divalent Sn) solutions consumes less electricity than alkali stannate solutions (tetravalent Sn) [42–45]. The major drawback of this system is Sn oxidation. At high current density, soluble Sn anodes are passivated due to the formation of SnO_2. Irregular, dendritic, needle-like electrodeposits of tin are in general obtained from acidic electrolytes without organic additives [47–49]. To improve the surface finish, morphology, and adhesion during acidic tin plating, various organic chemicals as additives have been investigated in the past [50–52]. In the literature, we observe that sulfate baths have been used widely to plate pure Sn and various Sn alloys in electronics industries [53].

4.3. Stannous chloride baths

For tin electroplating, sulfate baths use a number of additives to produce a homogeneous deposit [47–53]. However, in-spite of these advantages, the use of additives is undesirable for health concerns. Moreover, adverse effects on plating efficiency and the working environment have been observed [54, 55]. In case of alkaline baths, the same is true as it requires heating the solution that causes bubble formation [44–46]. Also tin is tetravent in alkaline baths causing more power. In view of these disadvantages, recently, He *et al.* developed a bath containing citric acid or its salt and an ammonium salt [53]. In the same line, Sharma *et al.* produced bulk tin films from stannous chloride-citrate bath and found improved thermal and mechanical properties [56].

4.4. Methanesulfonic acid baths

The problem of oxidation of Sn^{2+} to Sn^{4+} can be minimized by using a reducing acid Methanesulfonic acid (MSA) bath as compared to sulfate bath. MSA is much less corrosive to electronic materials than sulfuric acid [55]. It forms a clear solution, have high dissolving power and less sensitive to Sn oxidation at higher current densities [57]. Different combinations of methanesulfonate baths have been developed to electrodeposit both pure Sn and various Sn alloys by adding different additives [58].

4.5. Pyrophosphate baths

There are a number of pyrophosphate baths available in the literature [42]. These types of baths containing $P_2O_7^{2-}$ is considered as one of the most stable systems and is widely used for Sn plating [59, 60]. However, its use is limited in literature due to one disadvantage, that it requires more control and maintenance than the other plating baths. The operational temperature should not exceed 43–60°C because the pyrophosphate complex hydrolyzes to orthophosphate at temperatures higher than 60°C, which degrades the solution [42, 61].

5. Electrodeposition of Sn alloys

According to the electrochemical series of metals, we know that the reduction potential of two elements would be different [35]. For a solution containing two or more different metal ions at low overpotential, the metal with the most noble reduction potential will deposit at a faster rate [35, 36]. If the electrode potential difference is far apart, then metal alloy electrodeposition may be impossible. The difference can be eliminated in view of the Nernst equation by modifying the activity values and feasibility of the co-deposition can thus be determined [37, 39]. This can be achieved by inducing a considerable change in ionic concentrations via complex ion using certain complexing agents. In the past, thiourea has been utilized as a complexing agent for the electrodeposition of Sn-Ag and Sn-Ag-Cu alloys [62, 63]. The formation of complexes by bonding the metal ions with complexing agents will decrease the concentration of the free metal ions in the solution significantly and modify the reduction potential of the metal ions [64, 65]. With an increase in overpotential, the electrodeposition reaction will progress from the charge transfer region to the mix and then to the mass-transfer regime. Under mass transfer control at sufficiently higher overpotentials, the relative deposition rates of two or more metals will be governed by the concentrations and the diffusion coefficients of the metal ions [38, 39]. For alloy deposition, the basic mechanism remains the same as the Sn deposition discussed in the previous section. The summary of the previous studies on different types of plating baths are given in Table 1.

Coating	Processing Methods	Plating bath	Applications	Reference
Sn	Electroplating	Sodium stannate bath + halides	Lead-free	Abdel Rahim *et.al.*, 1986 [44]
Sn	Electroplating	Sodium stannate bath + sorbitol	Soldering	Broggi *et al.*, 2003 [45]
Sn	Pulse electroplating	Sodium stannate bath	Lead-free	Sharma *et al.*, 2012 [46]
Sn-Co	Electroplating	Stannous sulfate Gluconate bath	Lead-free	Rehim *et al.*, 1996 [48]
Sn	Electroplating	Stannous sulfate bath + aldehydes	Solder joints	Tzeng *et al.*, 1996 [49]
Sn	Electroplating	Stannous sulfate bath + Benzal acetone and N,N-bis(tetraoxyethylene)octad ecylamine	Lead-free	Nakamura *et al.*, 1997 [50]

Coating	Processing Methods	Plating bath	Applications	Reference
Sn-Cu, Sn-Bi, and Sn-Ag-Cu	Electroplating	Stannous sulfate bath + polyoxyethylene laurylether	Lead-free	Fukuda *et al.*, 2003 [51]
Sn-Co	Electroplating	Stannous sulfate bath + DS-10 synthanol, coumarin, formalin (37% solution)	Flip Chip technology	Medvedev *et al.*, 2001 [52]
Sn	Pulse electroplating	Stannous sulfate bath + triton X 100	Solder joints	Sharma *et al.*, 2013 [54]
Sn	Pulse electroplating	Stannous chloride bath	Wafer bumps	He et al., 2008 [53]
Sn	Pulse electroplating	Stannous chloride bath	Lead-free	Sharma *et al.*, 2013 [56]
Sn	Electroplating	MSA bath + PEG, PPG+ Phenolphthalein	Solder bumps	Martyak *et al.*, 2004 [57]
Sn	Electroplating	MSA bath + per-fluorinated cationic surfactant	Solders	Low *et al.*, 2008 [58]
Sn	Electroplating	Pyrophosphate bath + dextrin+ gelatin	Solders	Vaid *et al.*, 1957 [60]
Sn	Electroplating	Pyrophosphate bath + gelatin	Li-ion battery	Kim *et al.*, 2010 [59]
Sn-Ag and Sn-Ag-Cu	Electroplating	Stannous sulfate bath +thiourea	Solders	Ozga *et al.*, 2006 [62]
Sn-Ag-Cu	Electroplating	Stannous sulfate bath +thiourea+ polyoxyethylene laurylether	Lead-free	Fukuda *et al.*, 2002 [63]
Sn-Zn	Electroplating	Stannous chloride bath + PEG	Lead-free	Lin *et al.*, 2003 [64]
Ni, Sn, and Ni-Sn	Electroplating	Pyrophosphate-glycine bath	Lead-free	Lacnjevac *et al.*, 2012 [65]

Table 1. Summary of various studies on electrodeposition of Sn.

6. Process parameters and nanostructure formation

The electrodeposition method is dependent on several processing parameters involved in the electrodeposition process. Therefore, to obtain the desired properties, it is essential to optimize the operating parameters. The effect of these parameters on tin electrodeposition is discussed in the following sections.

6.1. Current density

The current density is the primary controlling parameter in pulse electrodeposition. The average current density $J_{average}$ is given by peak current density J_{peak} x Duty Cycle. This enables us to achieve a very high current density in pulse electrolysis without depleting the metal ions at the electrode surface [66]. A high overpotential associated with higher pulse current density raises the nucleation rate resulting in fine grained deposit and thus improved properties can be obtained [38–40]. For a stable growth, the critical radius of the electrodeposited nucleus varies inversely to the cathodic overpotential. The higher the cathodic overpotential, the higher the nucleation rate will be and vice versa [66, 67]. Thus, the grain size decrease with the increase in cathodic overpotential and finer grains are obtained. A higher current density also leads to the development of dendritic morphology of the deposits [49–54, 56]. The dendritic morphology can be suppressed by using additives. For alloy deposition, the change in alloy morphology has been noted by several researchers with increase in current densities [64, 65, 68].

The dendritic or irregular shaped morphology in response to the current density is due to the fact that if there is a significant increase in the current density, it increases the nucleation rate and is also associated with a higher rate of hydrogen evolution. Due to high overpotential at this state, the rate of diffusion of Sn2+ ions towards cathode becomes significant over its replenishment from the electrolyte [68, 69]. This phenomenon creates a concentration gradient in the vicinity of electrolyte/cathode interface. As a consequence, the deposition occurs preferentially on certain protrusions (heterogeneous active sites on the cathode in a random manner. This type of non-uniform growth kinetics has also been observed in copper plating [70]. The cathodic current efficiency at this stage is poor indicating a rapid hydrogen evolution which decreases the concentration of Sn2+ ions in the electrolyte.

6.2. pH

The effect of pH in tin plating has not been discussed in detail in the literature. The majority of the tin plating baths available in literature are acidic in nature. As a general statement the rate of deposition increases with pH [71]. Bath pH not only affects the deposition rate but modifies the crystal orientation. Some people have noted a pH induced texture in Sn grains from acidic electrolyte [72]. Ebrahimi *et al.* investigated the effect of pH on nickel deposits [73]. He found that the pH values of 4.5, 4.7, and 5.0 resulted in grain size of 79, 45, and 56 nm, respectively. The increasing pH value from 2.8 to 5.1 enhanced the nucleation rate of nickel crystals, which corresponded to decreasing grain size from 343 to 35 nm. For copper electrodeposition, Natter and Hempelmann reported the smallest copper grain size of 8 nm at pH 1.5–2.0 and a continuous increase of the grain size up to 100 nm at pH 11.5 [74]. The morphology of the deposited Sn coatings with different bath pH gets refined with an increase in pH. At a high value of pH around 2, irregular and porous deposits are observed due to the progressive increase in hydrogen evolution reaction. However, at a very high pH (=3), the deposits turn powdery due to severe hydrogen gas evolution [54].

6.3. Temperature

The majority of acid type formulations operate at room temperature, while alkaline tin baths need to be heated at higher optimum temperatures. As the temperature increases, the rate of deposition also increases [41, 54, 74]. The velocity (diffusion and migration) of the metal ions and inhibitor molecules are functions of the temperature. The viscosity of the electrolyte decreases at high temperature, therefore, the diffusion rate and the velocity of metal ions and inhibitor molecules are increased. Sahaym *et al.* studied the effect of bath temperature in Sn electrodeposits and found the pyramid type morphology at elevated temperatures [75]. They varied the bath temperature from 35°C to 85°C and explained that at higher bath/substrate temperature, the cooling rate of the adatoms will be slower. This indicates that at higher plating temperatures, adatoms can travel to longer diffusion distances on the cathode. This shows that an increase in the bath temperature may lead to increased grain growth and a smooth coating surface. Similar results are shown by Sharma *et al.* [74]. Bath temperature has been also shown to affect the morphology of the whiskers that formed upon aging at room temperature [75]. The metal ions and the inhibiting species have a higher mobility at high temperatures. It results in an increase in the concentration of metal ions towards the cathode and a decreases in the cathodic overpotential is observed. This increases the energy barrier (ΔG) for the nucleation process according to the equation given by Glasstone in reference [74], resulting in coarser grain size at higher temperature:

$$\Delta G \propto \left\{ \frac{1}{\left(\eta + \left[\dfrac{C'}{C} \right] \right)^2} \right\} \tag{13}$$

where C' is the activity of the Sn^{2+} on the electrode and C is the activity of Sn^{2+} in the bulk solution. The adsorption rate of inhibiting species decreases at high temperatures due to the decrease in viscosity causing an enhanced grain growth. It is reported in the past that in an acidic plating bath may decompose at high temperature [41, 42]. Therefore, the plating bath color may also change from light to dark yellow and get precipitated. This may be due to the decomposition of the sulfate entity or oxidation of Sn^{2+} to Sn^{4+}, which affects the solution chemistry [41–42, 45, 46]. The plating bath decomposition at elevated temperatures will decrease the bath conductivity and hence the current efficiency decreases beyond 40°C, with a negligible change in deposit morphology.

6.4. Additives

Although literature review about composite films and nanocrystalline films contains information about additives, additives for conventional metal deposition are still important as those additives would provide another aspect of information about deposition mechanism and novel additives. Popular organic additives that have been used in electrodeposition are gelatin, thiourea, EDTA (Ethylenediaminetetracetic acid), citric acid, benzotriazole (BTA), and inorganic additives such as chloride, are covered in this chapter. Nakamura *et al.* used various organic additives such as benzalacetone (BA) and N, N-bis (tetraoxyethylene) octadecylamine [50]. They found that by increasing the concentration of BA, smaller grains are obtained and roughness increases. Further, when both are used with a high concentration of BA, fine-grained deposits are obtained due to a synergistic action of these adsorbed species. Tzeng *et al.* studied the behavior of aliphatic and aromatic aldehyde groups in tin electrodeposition. They found that as compared to formaldehayde and propionoaldehyde, benzaldehyde is hydrophobic in nature showing more overpotential and thus acts as the best grain refiner [49]. Surface active agents such as Triton X-100, aromatic carbonyl compounds, amine-aldehyde reaction products, methane sulphonic acid and its derivatives, etc. [50, 52, 57, 58, 76, 77], are added to plating baths. The addition of excessive amounts of additives, high current density, and an increase in the concentration of metal ions in the electrolyte may affect the deposit solderability. It is mainly attributed to the oxidation of tin ions in the solution, as well as the aging of the deposit in the presence of organic brighteners [76–78]. Fine-grained and smooth deposits were obtained from acid stannous sulphate solutions containing some aromatic ketones [50]. It was found that these organic compounds were adsorbed on the cathode surface and enhanced the overpotential. In acidic solutions, many other chemicals, such as polyoxyethylene laurylether, triethanolamine, sorbitol, sodium gluconate, 1, 4- hydroxybenzene, Trion X-100, or polypropylene glycol may be added as complexing agents. The additives are necessary to improve solution stability and deposit morphology. However, an excessive amount of additives would make the coating less solderable [54, 57, 58, 78].

For alkaline stannate bath additives, few reports are available in literature. They usually do not require any additives since deposition occurs at a very negative potential and hydrogen evolution runs parallel with tin deposition acting as a leveler. However, the disadvantage is the diffusion of hydrogen inside the tin deposits [74–79]. In literature, the effect of additives has not been discussed in detail for chlorides electrolytes. Sekar *et al.* investigated the effect of additives in detail on pulse elctrodeposition of tin containing Gelatin, β-naphthol, polyethylene glycol, peptone, and histidine as additives [80]. They observed a grain refinement brought about by additives but a marked decrease in current efficiency and found that the bath containing peptone has the lowest crystallite size with better corrosion properties. The selection of organic additive(s) generally depends upon the nature of the metal ions, bath pH, as well as the temperature of the plating solution/substrate. For instance, semi-bright or bright nickel plating solution has different sets of additives as compared to the additives used in acidic zinc plating. Similarly, additives used in acid tin plating baths are different than those used in alkaline tin plating. However, the coating morphologies and properties obtained from acidic and alkaline solutions may or may not be similar to each other. In MSA baths, glycol type additives, PEG, PPG, and Phenolphthalein have been tried in literature [55]. Glycol-type additives are effective in minimizing hydrogen gas evolution at low overpotentials and refining the grain structure across a wide current density range. These additives are active in the presence of Phenolphthalein [57]. Low and Walsh used the perfluorinated surfactant in MSA bath and observed that the adsorption of the surfactant hindered hydrogen evolution and reduced the peak and limiting current densities [58]. There is little information on pyrophosphate baths in literature. Pyrophosphate baths include mostly glycol additives used in SnAg type solder bumps. These baths require more control of the electrodeposition parameters. Normally chelating agents and organic additives such as triethanolamine, sodium gluconate, 1, 4-hydroxybenzene, and Triton X-100 have been added in pyrophosphate baths [42, 59, 60, 61, 81, 82]. The additives are active only up to a certain concentration and have a beneficial effect on the microstructure of the deposits by increasing the cathodic polarization.

At higher concentration of additives, the progressive evolution of hydrogen gas leads to the development of non-uniform powdery deposits. The powdery deposits generally arise due to the adsorption and absorption of the hydrogen gases in the deposits according to the following reactions [54, 83]:

$$2H_{ads} \text{---} H_2 \tag{14}$$

$$H^+ + e^- \text{----} H_{ads} \tag{15}$$

$$H_{ads} \text{----} H_{abs} \tag{16}$$

The additive blocks the association of generated hydrogen atoms according to Equation (14). Consequently, the concentrations of H_{ads} rises following Equation (15). The combined effect of these two phenomena (Equations 15 and 16) results in absorption of hydrogen atoms in the deposits according to Equation (16).

6.5. Duty cycle and frequency

The pulse deposition rate is given by the pulse current density and other parameters such as 'on' time (T_{on}) and 'off' (T_{off}) time. The pulse length T_{on} and T_{off} time have significant effect on grain size. An increase in pulse current density at constant length T_{on} and constant average current density with an increase in T_{off} time cause an increase in grain-size, because grain growth takes place during the T_{off} time. But an increase in pulsed current density at constant pulse charge and at constant T_{off} implies a decrease in grain-size. Therefore, a short T_{on} prevents grain growth and increases the nucleation rate [32].

The frequency of the pulse is the reciprocal of the total pulse duration consisting of T_{on} and T_{off}. The pulse length T_{on} and T_{off} time had significant effect on grain size. A higher pulse duration at lower pulse frequencies indicates enough time for the charging (Ton) and discharging $(Toff)$ of the double layer [36–38]. In contrast, at a high pulse frequency the double layer does not have enough time for charging and discharging [39, 40]. Thus, lower frequencies cause bigger grains while higher frequencies results in a grain refinement of the electrodeposits [36–41, 84].

The deposition rate in the pulse technique is governed by the pulse current density (J_{peak}), and on time (T_{on}) and off time (T_{off}). At a given peak current density, the duty cycle is given by:

$$\text{Duty Cycle } (\%) = \frac{T_{on}}{T_{on} + T_{off}} \qquad (17)$$

As discussed already, the average current density ($J_{average}$) in a pulse electrodeposition is given by:

$$J_{average} = \text{Duty Cycle} \times J_{peak} \qquad (18)$$

where J_{peak} is the peak current density. A 100% duty cycle means there is no pulse used. The effect of pulse parameters has been studied in a paper by Sharma *et al.* [54], where they used various combinations of pulse on and on time to produce several duty cycles as shown in Table 2. These combinations are not unique and can be varied according to the desired changes in deposit microstructures.

Duty Cycle (%)	T_{on} (s)	T_{off} (s)
4	0.00042	0.01
10	0.0011	0.01
20	0.0025	0.01
40	0.0067	0.01
60	0.0150	0.01

(Reprinted "With kind permission from Springer Science+Business Media: Journal of Metallurgical and Materials Transactions A, Volume 45, 2014, Issue 10, Page 4610-4622, A. Sharma, S. Bhattacharya, S. Das, K. Das, Table II.

Table 2. Pulse parameters at various duty cycles.

However, at a duty cycle of 100%, the grain size observed is usually much coarser. Thus, it is inferred that the grain size of the deposits decreases with an increase in *Ton* (at constant T_{off} and J_{peak}). An increase in *Ton* (longer current on times) results in an increased overpotential [84]. Youssef and his co-workers have also observed similar behavior in zinc electrodeposits [85]. It is reported that the deposit grain size decreases up to a duty cycle of 44%, and an increase in the grain size is noticed beyond duty cycle of 44%. This may be explained due to the fact that the pulse deposition gets transformed to the direct current deposition at this stage [85].

The pulse current (PC) with lower duty cycle (<20%) produces uniform and compact deposits. At higher duty cycles (>20%) and at direct current (DC) deposition, severe increase porosity is observed. This may be due to higher average current flow time *Ton*, which triggers hydrogen gas evolution [86]. This type of porous deposit morphology can be attributed to the dissolved hydrogen and oxygen gases in the plating bath. A higher solubility of hydrogen (1617 and 43.3

mg/L in water) compared to that of the oxygen in water (43.3 mg/L), at 25ºC and a 1 bar favors the dissolution of hydrogen gas and its incorporation over oxygen in the deposits [86]. Therefore, the morphology is rather porous and rough beyond a duty cycle of 20%.

A reduced porosity in case of PC deposition with lower duty cycles can be correlated to the two factors. (i) partial diffusion of the hydrogen and oxygen gas away from the substrate during off time (T_{off}) and hence a suppressed absorption of hydrogen gas in the deposit; (ii) the total amount of gas generated during the electrolysis of water in on time (T_{on}) is less compared to a continuous DC [54, 86]. Therefore, pulse current with lower duty cycle will result in a uniform and dense morphology as compared to high duty cycles or DC deposition.

The morphology of electrodeposits is also influenced by the pulse frequency in the electrode-position. The pulse frequency parameters varied according to Sharma *et al.* [54] and are shown in Table 3. These combinations are also not unique like the duty cycle and can be varied to obtain the optimum deposit microstructures.

Frequency (*f*) of the pulse is described as follows:

$$f = \frac{1}{\left(T_{on} + T_{off}\right)} \tag{19}$$

Frequency	10 Hz		50 Hz		100 Hz		500 Hz	
Duty Cycle	T_{on}(s)	T_{off}(s)	T_{on}(s)	T_{off}(s)	T_{on}(s)	T_{off}(s)	T_{on}(s)	T_{off}(s)
10%	0.01	0.09	0.002	0.018	0.001	0.009	.0002	.0018

(Reprinted "With kind permission from Springer Science+Business Media: Journal of Metallurgical and Materials Transactions A, Volume 45, 2014, Issue 10, Page 4610-4622, A. Sharma, S. Bhattacharya, S. Das, K. Das, Table III.

Table 3. Pulse frequency parameters

Lower pulse frequency (*f*) indicates a longer pulse period ($T_{on}+T_{off}$) at a constant duty cycle (D), as shown in Equation (19). The pulse duration is higher at lower frequencies providing sufficient time for the charging and discharging of the double layer.

The Sn atoms can migrate freely to the most stable position facilitating the grain growth. An increase in the pulse frequency shortens the pulse duration, i.e., both T_{on} and T_{off} are smaller as shown in Table 3. Shanthi *et al.* have noticed a severe grain refinement during pulse current plating of silver alloy with short pulse and high frequency [84]. As already discussed, at high pulse frequencies the charging an discharging action of double layer gets poor. These phenomena results in very thin diffusion layers that make the transportation of the migrating Sn ions towards cathode difficult. The nucleation rate improves as a consequence a dense microstructure is obtained [85, 66].

6.6. Effect of agitation

Agitation in the plating solution can be produced either by agitating the electrolyte or by moving the cathode. At low agitation rates, the effect of agitation on deposit composition is not visible, while the agitation rates may decrease the tin content in the coatings. Moreover, agitation may also increase the deposit roughness up to some extent [42, 54]. Agitation has beneficial effects of increasing the plating rate and permits the use of higher current densities by lowering polarization [36, 37, 38, 39]. Wen and Szpunar studied the nucleation and growth of tin and pointed out that agitation should not exceed beyond a certain limit where turbulent flow occurs that cause the difficulty of tin ions supply to the cathode even at high current densities [87]. Thus the cathode coverage is poor and the deposition rate decreases.

It is interesting to note that the stirring rate of the bath has a significant effect on the deposit morphology. The deposition parameters, except the stirring speed that controls the bath agitation, are kept constant. During still deposition, the cathode coverage is poor, and irregular and non-uniform deposits are obtained. This can be explained as when no agitation is provided, the depositing ions from the electrolyte get deposited preferentially on the cathode. Thus, a concentration gradient is established in the vicinity of electrolyte/cathode interface. The deposition is uneven at this stage and morphology is very poor. When the bath stirring is increased, the concentration gradient is decreased and deposition rate increases [87, 88]. Further stirring of the bath will cause fast transportation of metal ions towards cathode. At a sufficiently high stirring rates of the plating bath, the flow of the electrolyte will be turbulent and the metal ions may move away from the cathode vigorously and a lowering in the deposition rate is observed [88].

7. Sensitivity of the variables and grain size distribution

The electrodeposition parameters considered in this investigation include current density, concentration of the additive, duty cycle, frequency, pH, temperature, and agitation. The obtained results in this work indicate that the pulsed current electrodeposition can be an efficient method for the electrosynthesis of tin deposits. The surface morphology evolution depends on the electrodeposition parameter that tries to modify the overpotential, in a direct or indirect way. The current density is the most sensitive of all the plating parameters which affects the deposit morphology severely. The nucleation rate and grain growth can be significantly controlled by changing the duty cycle to lower values up 5 to 20%. At higher duty cycles, the porous deposits are produced.

Smaller pulse frequency gives large grained deposits. Thus, a combination of duty cycle and pulse frequency can be optimized for an ultrafine grained or an optimum grain morphology depending upon the application. The presence of additives in the plating bath improves the surface finish and morphology if it is added up to its optimum concentration. The current density is found to decrease with bath pH increase. The grain size decreases as pH value increases due to the increase in the cathodic polarization. It is also noteworthy point that powdery deposits are too developed at very high pH due to the precipitation of stannic hydroxides. An increase in bath temperature is noticed to raise the grain size of the deposits. However, a decrease in the electrolyte conductivity and current efficiency is noted at elevated temperatures due to the precipitation of metal ions in the plating bath. Bath stirring improves the availability of metal ions towards cathode and thus the deposition rate is enhanced before the flow of electrolyte turns turbulent where the metal ions move away from the cathode. The grain size is also increased due to the decrease in overpotential with bath stirring rate

Author details

Ashutosh Sharma[1*], Siddhartha Das[2] and Karabi Das[2]

1 Department of Materials Science and Engineering, University of Seoul, Seoul, Korea

2 Department of Metallurgical and Materials Engineering, Indian Institute of Technology Kharagpur, Kharagpur, India

References

[1] Abtew M, Selvaduray, G. Lead Free Solders in Microelectronics. Materials Science and Engineering R. 2000;27:95-141.

[2] Suganuma, K. Lead Free Soldering in Electronics – Science, Technology and Environmental Impact. New York: Marcell Dekker Inc; 2004.

[3] Tu KN, Zeng K. Tin-Lead (SnPb) Solder Reaction in Flip Chip Technology. Materials Science and Engineering R. 2001;34:1-58.

[4] Tummala RR. Fundamentals of Microsystems Packaging. McGraw-Hill; 2000. p. 185-210.

[5] Liu JP, Guo F, Yan YF, Wang WB, Shi YW. Development of Creep-Resistant, Nano-sized Ag Particle-Reinforced Sn-Pb Composite Solders. Journal of Electronic Materials. 2004;33(9):958-963.

[6] Sobiech M, Teufel J, Welzel U, Mittemeijer EJ, Hugel W. Stress Relaxation Mechanisms of Sn and SnPb Coatings Electrodeposited on Cu: Avoidance of Whiskering. Journal of Electronic Materials. 2011;40:2300-2313.

[7] Coombs C. Coombs' Printed Circuits Handbook. USA: McGraw Hill; 2001. p. 45.1-45.9.

[8] Subramanian KN. Lead free electronic solders. A special issue of the Journal of Materials Science: Materials in Electronics. New York: Springer science; 2007. p. 55-76.

[9] Suganuma K. Advances in Lead-Free Electronics Soldering. Current Opinion in Solid State and Materials Science. 2001;5:55-64.

[10] Puttlitz KJ, Gaylon GT. Impact of RoHS Directive on High Performance Electronic Systems. Journal of Materials Science: Materials in electronics. 2007;18: 347-365.

[11] Gilleo K. Area Array Packaging Handbook. New York: McGraw-Hill; 2002.

[12] Puttiltz KJ, Stalter KA. Handbook of Solder Technology for Microelectronic Assemblies. New York: Marcell Dekker Inc; 2004. p. 51-55.

[13] Pecht M, Fukuda Y, Rajagopal S. The Impact of Lead-Free Legislation Exemptions on the Electronics Industry. IEEE Transactions on electronics packaging manufacturing. 2004;27:221-232.

[14] Plumbridge WJ. Tin Pest Issues in Lead-Free Electronic Solders. Journal of Materials Science: Materials in Electronics. 2007;18:307–318.

[15] Wang Y, Lu KH, Gupta V, Stiborek L, Shirley D, Im J, Ho PS. Effect of Sn Grain Structure on Electromigration Reliability of Pb-Free Solders. In: Proceedings of Electronic Components and Technology Conference; 2011. p. 711-716.

[16] Alam ME, Nai SML, Gupta M. Development of High Strength Sn–Cu Solder Using Copper Particles at Nanolength Scale. Journal of Alloys and Compounds. 2009;476:199-206.

[17] Shen J, Chan YC. Research Advances in Nano-Composite Solders. Microelectronics Reliability. 2009;49:223-234.

[18] Amagai M. A Study of Nanoparticles in Sn–Ag Based Lead Free Solders. Microelectronics Reliability. 2008;48:1-16.

[19] Chen Z, Shi Y, Xia Z, Yan Y. Properties of Lead-Free Solder SnAgCu Containing Minute Amounts of Rare Earth. Journal of Electronic materials. 2003;32(4):235-243.

[20] Brusse JA, Ewell GJ, Siplon JP. Tin Whiskers: Attributes And Mitigation. In: Proceedings of 16th Passive Components Symposium (CARTS EUROPE 2002); 2002. p. 221-233.

[21] Tu KN. Cu-Sn Interfacial Reactions: Thin-Film Case Versus Bulk Case, Materials Chemistry and Physics. 1996;46:217-223.

[22] Takenaka T, Kano S, Kajihara M, Kurokawa N, Sakamoto K. Growth Behavior of Compound Layers in Sn/Cu/Sn Diffusion Couples During Annealing at 433–473K. Materials Science and Engineering. 2005;396:115-123.

[23] Nakahara S, McCoy RJ, Buene L, Vandenberg JM. Room Temperature Interdiffusion Studies of Au/Sn Thin Film Couples. Thin solid films. 1981;84:185-196.

[24] Simic V, Marinkovic Z. Room Temperature Interactions in Copper Metal Thin Film Couples. Journal of less common metals. 1980;72:133-140.

[25] Tang WM, He A, Liu Q, Ivey DG. Solid State Interfacial Reactions in Electrodeposited Ni/Sn Couples. International Journal of Minerals, Metallurgy and Materials. 2010;17(4):459-463.

[26] Marinkovic Z, Simic V. Room Temperature Interactions in Ni/Metal Thin Film Couples. Thin solid films. 1982;98:95-100.

[27] Gaylon GT. Annotated Tin Whisker Bibliography. NEMI Tin Whisker Modeling Project; July 2003.

[28] Liangfeng L, Tai Q, Jian Y, Yongbao F. Synthesis of Ag-Cu-Sn Nanocrystalline Alloys as Intermediate Temperature Solder by High Energy Ball Milling. Advanced Materials Research. 2009;79-82:449-452.

[29] Hong Z, Ming TW, Qing XG, Cheng WY, Xiang ZZ. Synthesis of Sn–Ag Binary Alloy Powders by Mechanical Alloying. Materials Chemistry and Physics. 2010;122: 64-68.

[30] Aggarwal AO, Abothu IR, Raj PM, Sacks MD, Tummala RR. Lead-Free Solder Films Via Novel Solution Synthesis Routes. IEEE Transactions on Components and Packaging Technologies. 2007;30:486-493.

[31] Hao H, Tian J, Shi YW, Lei YP, Xia ZD. Properties of Sn3.8Ag0.7Cu Solder Alloy with Trace Rare Earth Element Y Additions. Journal of Electronic Materials. 2007; 36 (7): 766-774.

[32] Hsiao Li-Yin, Duh Jenq-Gong. Synthesis and Characterization of Lead Free Solders With Sn-3.5Ag-xCu (x = 0.2, 0.5, 1.0) Alloy Nanoparticles by the Chemical Reduction Method. Journal of Electrochemical Society. 2005;159(9):J102-J109.

[33] Conway PP, Fu EKY, Williams K. Precision High Temperature Lead-Free Solder Interconnections by Means of High-Energy Droplet Deposition Techniques. CIRP Annals–Manufacturing Technology. 2002;51:177-180.

[34] Hsiung CK, Chang CA, Tzeng ZH, Ho CS, Chien FL. Study on Sn-2.3Ag Electroplated Solder Bump Properties Fabricated by Different Plating and Reflow Conditions. In: Proceedings of 9th Electronics Packaging Technology Conference; 2007. p. 719-724.

[35] Ruythooren W, Attenborough K, Beerten S, Merken P, Fransaer J, Beyne E, Hoof CV, Boeck JD, Celis JP. Electrodeposition for the Synthesis of Microsystems. 2000;10:101-107.

[36] Paunovic M, Schlesinger M. Fundamentals of Electrochemical deposition. New Jersey: John wiley and Sons; 2006.

[37] Brown R. RF/Microwave Hybrids Basics. Materials and Processes. Dordrecht: Kluwer Academic Publishers; 2004. p. 169-171.

[38] Bard AJ, Faulkner LR. Electrochemical Methods: Fundamentals and Applications. New York: John Wiley & Sons; 2001.

[39] Kanani N. Electroplating: Basic Principles, Processes and Practice. Oxford: Elsevier Ltd; 2004.

[40] Chandrasekar MS, Pushpavanam M. Pulse and Pulses Reverse Plating-Conceptual, Advantages and Applications. Electrochimica Acta. 2008;53:3313-3322.

[41] Djokic SS. Modern Aspects of Electrochemistry-Electrodeposition Theory and Practice. New York: Springer; 2010.

[42] Tan AC. Tin and Solder Plating in the Semiconductor Industry. London: Chapman &. Hall; 1993.

[43] Schlesinger M, Paunovic M. Modern Electroplating. Hoboken, New Jersey: John wiley and sons; 2010.

[44] Abdel Rahim SS, Awad A, El Sayed A. Role of halides in the electroplating of tin from the alkaline stannate bath. Surface and coating Technology. 1986;28:139-150.

[45] Broggi RL, Oliviera GMD. Barbosa LL, Pallone EMJA, Carlos IA. Study of an Alkaline Bath for Tin Deposition in the Presence of Sorbitol and Physical and Morphological Characterization of Tin Film. Journal of Applied Electrochemistry. 2006;36:403-409.

[46] Sharma A, Bhattacharya S, Sen R, Reddy BSB, Fecht H-J, Das K, Das S. Influence of Current Density on Microstructure of Pulse Electrodeposited Tin Coatings. Materials Characterization. 2012;68:22-32.

[47] Lowenheim F A. Modern Electroplating. New York: John Wiley and Sons; 1974.

[48] Rehim SSAE, Refaey SA, Schwitzgebel G, Taha F, Saleh MB. Electrodeposition of Sn-Co Alloys From Gluconate Baths. Journal of applied electrochemistry. 1996;26:413-418.

[49] Tzeng GS, Lin SH, Wang YY, Wan CC. Effect of Additives on the Electrodeposition of Tin From an Acidic Sn(II) Bath. Journal of applied electrochemistry. 1996;26:419-423.

[50] Nakamura Y, Kaneko N, Nezu H. Surface Morphology and Crystal Orientation of Electrodeposited Tin From Acid Stannous Sulphate Solutions Containing Various Additives. Journal of applied electrochemistry. 1994;24:569-574.

[51] Fukuda M, Imayoshi K, Matsumoto Y. Effect of Adsorption of Polyoxyethylene Laurylether on Electrodeposition of Pb-free Sn Alloys. Surface and Coatings Technology. 2003;169-170:128-131.

[52] Medvedev GI, Makrushin NA. Electrodeposition of Tin From Sulfate Electrolyte With Organic Additives. Russian Journal of Applied Chemistry. 2001;74(11):1842-1845.

[53] He A, Liu Q, Ivey DG. Electrodeposition of Tin: A Simple Approach. Journal of Materials Science: Materials in Electronics. 2008;19:553-562.

[54] Sharma A, Bhattacharya S, Das S, Das K. A Study on the Effect of Pulse Electrodeposition Parameters on the Morphology of Pure Tin Coatings. Metallurgical and Materials Transactions A. 2013;45A:4610-4622.

[55] Gillman HD, Fernandes B, Wikiel K. Metal Alloy Sulfonic Acid Electroplating Baths. US Patent 6183 619 B1; Feb 6 2001.

[56] Sharma A, Bhattacharya S, Das S, Das K. Influence of current density on surface morphology and properties of pulse plated tin films from citrate electrolyte. Applied Surface Science. 2014;290:373-380.

[57] Martyak NM, Seefeldt R. Additive-Effects During Plating in Acid Tin Methanesulfonate Electrolytes. Electrochimica Acta. 2004;49:4303-4311.

[58] Low CTJ, Walsh FC. The Influence of a Perfluorinated Cationic Surfactant on the Electrodeposition of Tin from a Methanesulfonic Acid Bath. Journal of Electroanalytical Chemistry. 2008;615:91-102.

[59] Kim RH, Nam DH, Kwon HS. Electrochemical Performance of a Tin Electrodeposit With a Multi-Layered Structure for Li-ion Batteries. Journal of Power Sources. 2010;195:5067-5070.

[60] Vaid J, Char TLR. Tin Plating From the Pyrophosphate Bath. Journal of Electrochemical society. 1957;104(5):282-287.

[61] Harper CA. Electronic Materials and Processes Handbook. New York: McGraw-Hill; 2004.

[62] Ozga P. Electrodeposition of Sn-Ag and Sn-Ag-Cu Alloys From Thiourea Aqueous Solutions. Archives of Metallurgy and Materials. 2006;51:413-421.

[63] Fukuda M, Imayoshi K, Matsumoto Y. Effects of Thiourea and Polyoxyethylene Lauryl Ether on Electrodeposition of Sn-Ag-Cu Alloy as a Pb-Free Solder. Journal of Electrochemical Society. 2002;149:C244-C249.

[64] Lin KL, Sun LM. Electrodeposition of eutectic Sn–Zn alloy by pulse plating. Journal of Materials Research. 2003;18:2203-2207.

[65] Lačnjevac U, Jović BM, Jović VD. Electrodeposition of Ni, Sn and Ni–Sn Alloy Coatings from Pyrophosphate-Glycine Bath. Journal of Electrochemical Society. 2012;159(5):D310-D318.

[66] Ibl N. Some Theoretical Aspects of Pulse Electrolysis. Surface Technology. 1980;10:81-104.

[67] Watanabe T. Nano-Plating Microstructure Control Theory of Plated Film and Data Base of Plated Film Microstructure. Oxford UK: Elsevier Ltd; 2004.

[68] Sharma A, Bhattacharya S, Das S, Das K. Fabrication of Sn-Ag/CeO2 Electro-Composite Solder by Pulse Electrodeposition. Metallurgical and Materials Transactions A. 2013;44A:5587-5601.

[69] Liu Z, Zheng M, Hilty RD, West AC. Effect of Morphology and Hydrogen Evolution on Porosity of Electroplated Cobalt Hard Gold. Journal of Electrochemical Society. 2010;157 (7):D411-D416.

[70] Nikolic ND, Popov KI, Pavlovic LJ, Pavlovic MG. Morphologies of Copper Deposits Obtained by the Electrodeposition at High Overpotentials. Surface and Coatings Technology. 2006;201:560-566.

[71] Durney LJ. Electroplating Engineering Handbook. 4th ed. London: Chapman & Hall; 1984.

[72] Teshigawara T, Nakata T, Inoue K, Watanabe T. Microstructure of Pure Tin Electrodeposited Films, Scripta Materialia. 2001;44:2285-2289.

[73] Ebrahimi F, Bourne GR, Kelly MS, Matthews TE. Mechanical Properties of Nanocrystalline Nickel Produced by Electrodeposition. Nanostructured Materials. 2009;11(3): 343-350.

[74] Natter H, Hempelmann R. Nanocrystalline Copper by Pulsed Electrodeposition: The Effects of Organic Additives, Bath Temperature, and pH. Journal of Physical Chemistry. 1996;100:19525-19532.

[75] Sahaym U, Miller SL, Norton MG. Effect of Plating Temperature on Sn Surface Morphology. Materials Letters. 2010;64:1547-1550.

[76] Lee JY, Kim JW, Chang BY, Kim HT, Park SM. Effects of Ethoxylated α-Naphtholsulfonic Acid on Tin Electroplating at Iron Electrodes. Journal of The Electrochemical Society. 2004;151(5):C333-C341.

[77] Aragon A, Figueroa MG, Gana RE, Zagal JH. Effect of Polyethoxylate Surfactant on the Electrodeposition of Tin. Journal of Applied Electrochemistry. 1992;22: 558-562.

[78] Lal S, Moyer TD. Role of Intrinsic Stresses in the Phenomena of Tin Whiskers in Electrical Connectors. IEEE Transactions on Electronics Packaging Manufacturing. 2005;28:63-74.

[79] Song JY, Yu J, Lee TY. Effects of Reactive Diffusion on Stress Evolution in Cu–Sn Films. Scripta Materialia. 2004;51:167-170.

[80] Sekar R, Eagammai C, Jayakrishnan S. Effect of Additives on Electrodeposition of Tin and its Structural and Corrosion Behaviour. Journal of Applied Electrochemistry. 2010;40:49-57.

[81] Neveu B, Lallemand F, Poupon G. Mekhalif Z. Electrodeposition of Pb-free Sn alloys in pulsed current. Applied Surface Science. 2006;252:3561-3573.

[82] Correia AN, Facanha MX, Lima-Neto P. de Cu–Sn Coatings Obtained From Pyrophosphate-Based Electrolytes. Surface and Coatings Technology. 2007;201:7216- 7221.

[83] Franklin TC. Some Mechanisms of Action of Additives in Electrodeposition Processes. Surface and Coatings Technology. 1987;30:415-428.

[84] Shanthi C, Barathan S, Jaiswal R, Arunachalam RM, Mohan S. The Effect of Pulse Parameters in Electrodeposition of Silver Alloy. Materials Letters. 2008;62:4519- 4521.

[85] Youssef Kh MS, Koch CC, Fedkiw PS. Influence of Additives and Pulse Electrodeposition Parameters on Production of Nanocrystalline Zinc from Zinc Chloride Electrolytes. Journal of Electrochemical Society. 2004;151(2):C103-C111.

[86] Besra L, Uchikoshi T, Suzuki TS, Sakka Y. Application of Constant Current Pulse to Suppress Bubble Incorporation and Control Deposit Morphology During Aqueous Electrophoretic Deposition (EPD). Journal of the European Ceramic Society.

[87] 2009;29:1837-1845.

[88] Wen S, Szpunar JA. Nucleation and Growth of Tin on Low Carbon Steel. Electrochimica Acta. 2005;50:2393-2399.

[89] Musa AY, Slaiman QJM, Kadhum AAH, Takriff MS. Effects of Agitation, Current Density and Cyanide Concentration on Cu-Zn Alloy Electroplating. European Journal of Scientific Research. 2008;22(4):517-524.

7

A Process for Preparing High Graphene Sheet Content Carbon Materials from Biochar Materials

Yan-Jia Liou and Wu-Jang Huang

Abstract

Graphene is monolayer graphite and has higher electron mobility than silicon, high heat conduction, and special optical properties. In this study, we have attempted to use a two-step process (an acid pretreatment followed by a heat treatment) for producing high graphene sheet content (>80%) carbon materials (GSCCMs) from monocotyledonous and dicotyledonous biochar materials prepared at 350°C. The highest graphene sheet content of 83.86% is found with the CH_3COOH pretreatment followed by a 1500°C heat treatment of monocotyledonous biochar materials, and its conductivity was measured at 84.69 S/cm. Therefore, preparing GSCCMs from biochar materials could highly reduce the cost.

Keywords: Fullerenes, oxides, biomaterials, IR

1. Introduction

Graphene is monolayer graphite and has higher electron mobility than silicon, high heat conduction, and special optical properties. Graphene is composed of two-dimensional and hexagonal rings of sp^2 carbon atoms. The potential graphene application includes photoelectric elements, various medical applications, adsorbents for environmental pollutants, and use as a semiconductor material. The researchers used graphene as a transparent conductive oxide thin film. The results show that the electron and hole mobilities of the device are 1100 ± 70 cm^2/Vs and 550 ± 50 cm^2/Vs at the drain bias of –0.75 V, respectively [1]. In 2008, Hong et al. [2] mixed graphene and polyethylene dioxythiophene doped with polystyrene sulfonic acid

(PEDOT-PSS) to make dye-sensitized solar cells (DSSC). The results show that the optimal efficiency of photoelectric conversion is 4.5%. Hollanda et al. [3] mixed graphene and CNTs to make a nanocomposite as a gene transfection. The results show that the graphene/CNT nanocomposite has moderate to low cytotoxicity, high-transfection efficiency, and great potential as a gene carrier agent in nonviral-based therapy. In 2014, Sheshmani et al. [4] used magnetic graphene/chitosan for the removal of acid orange 7. From the results, it was found that the magnetic graphene/chitosan can effectively remove the anionic azo dyes from wastewater. Researchers' evaluations of graphene as an adsorbent for solid phase extraction with chlorophenols when applied to the analysis of river water samples had recoveries above 77.2% [5].

Many manufacturing methods of graphene have been proposed, such as chemical vapor deposition [6], chemical reduction of graphene oxide [7, 8], and exfoliation method [9]. In several studies, researchers have used a combined ozone–oxidation with annealing at 530 K to produce graphene powders from highly oriented pyrolytic graphite [10]. In 2010, Akhavan [11] used graphene oxide prepared graphene thin film through heat treatment. The results show that it can form a thickness of graphene thin film of 0.37 nm at 1000°C.

However, these processes are complicated by a high cost and the difficulty of removing by-products. Therefore, the market mechanisms of graphene have not yet been established at present. The authors have always used a low temperature and two-step pyrolized process in the production of graphene content carbon material from woody wastes. The first step is the charring process of biomass into biochar; the secondary step is to form graphene microcrystals by heating up to 1400°C. The results have shown that the most used secondary pyrolysis temperature at 1400°C has made graphene content carbon materials; the content of the graphene sheet was measured at 62.47%. In another work, we have found that the production of graphene content carbon from woody materials created a surface Raman quenching phenomenon [12, 13] (see Figures S2 and S3). In this study, we would like to report a modified process for directly preparing high graphene sheet content carbon materials (GSCCMs) from biochar materials.

2. Experimental

2.1. Preparation of biochar materials

In this study, biomass materials were sampled from the monocotyledonous materials (such as *Cocos nucifera, Elaeis guineensis Jacq*, and *Kingia australis*) and dicotyledonous materials (such as *Cinnamomum camphora, Cryptomeria japonica*, and *Aextoxicon punctatum*). The dicotyledonous biomass (*C. japonica*) was dried and cut into smaller pieces with dimensions of $5 \times 2 \times 4$ mm^3 in length, width, and height, respectively. The monocotyledonous biomass (*E. guineensis Jacq.*) was cut into smaller pieces with dimensions of $2 \times 0.05\text{–}0.2 \times 0.05\text{–}0.2$ mm^3 in length, width, and height, respectively. The biomass materials were heated at 350°C for 1 h to convert them into biochar materials. *E. guineensis Jacq.* and *C. japonica* were selected due to there are common woody materials in Taiwan.

2.2. Preparation of Graphene Sheet Content Carbon Materials (GSCCMs)

A two-step process with an acid pretreatment followed by a heat treatment produced high graphene sheet content carbon materials (GSCCMs) from monocotyledonous and dicotyledonous biochar materials. In the first acid pretreatment step, the above-mentioned biochar materials were immersed in 5 M of sulfuric acid (H_2SO_4) solution and 5 M of acetic acid (CH_3COOH) solution at room temperature for 14 days, respectively. After acid pretreatment, biochars were dried at $105 \pm 5°C$ before they were heated at 500°C, 800°C, 1200°C, and 1500°C under a vacuum.

The IDs of dicotyledonous samples are denoted as SC500, SC800, SC1200, and SC1500 (H_2SO_4 pretreatment), as well as AC500, AC800, AC1200, and AC1500 (CH_3COOH pretreatment), respectively. The IDs of monocotyledonous samples are denoted as SE500, SE800, SE1200, and SE1500 (H_2SO_4 pretreatment), as well as AE500, AE800, AE1200, and AE1500 (CH_3COOH pretreatment), respectively.

2.3. Property analysis of samples

An S-3000N (HITACHI, Japan) scanning electron microscope meter equipped with an energy dispersive spectrometer (SEM/EDX) was applied to analyze the structural exterior of the prepared GSCCM products. A Vector22 (Bruker, Germany) Fourier transform infrared spectrometer (FT-IR) was used to analyze their functional groups. A D8 Advance (Bruker, Germany) X-ray diffractometer (XRD) was used to analyze the lattice structures, and an FPSR-100 (Solar Energy Tech, Taiwan) four-point sheet resistance meter was used to analyze the samples' resistivity and conductivity.

3. Results and discussion

Figures 1 to 3 show the SEM image of GSCCM samples; surface erosion and the formation of graphite spherical grains are observed for biochar materials after acid pretreatment and heat treatment, respectively. The temperature necessary to generate a large number of graphite spherical grains was found to begin at 800°C. When the heating temperature was raised to 1500°C, the size of graphite spherical grains is increased. The authors' other publication had found that sodium ions have more contribution on the formation of those graphite spherical grains [14] (see Figure S4).

The FT-IR spectra of acid-pretreated dicotyledonous biochar materials are shown in Figure 4a, and those of monocotyledonous biochar materials are shown in Figure 4b. Both have amide groups (~3850 and 3745 cm^{-1}), aromatic groups (~ 878, 650 and 590 cm^{-1}), –OH (~3500 cm^{-1}), –CH_2- (~2950 cm^{-1}), C=O (~1630 cm^{-1}), –CH_3 (~1397 cm^{-1}), and C-O (~1100 cm^{-1}) stretching vibration motion. The peaks are from the cellulose and lignin structure of plants and corresponded to the peak of absorbed CO_2 molecules at 2250 cm^{-1}. After H_2SO_4 and CH_3COOH pretreatment, the peak at 1600 cm^{-1} split into outside-C=O and inside-C=O stretching vibration motion in the carbon layers, the peak of –CH_3 disappeared, and the peak at 1700 cm^{-1}

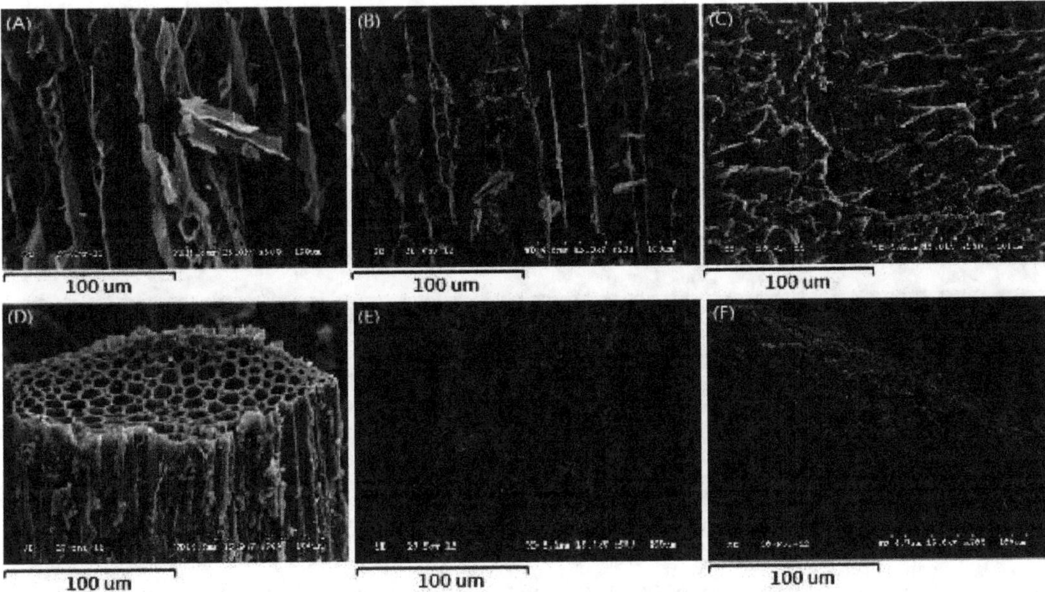

Figure 1. SEM image of (A) dicotyledonous biochar and after acid pretreatment for (B) H_2SO_4 and (C) CH_3COOH and (D) monocotyledonous biochar and after (E) H_2SO_4 and (F) CH_3COOH pretreatment.

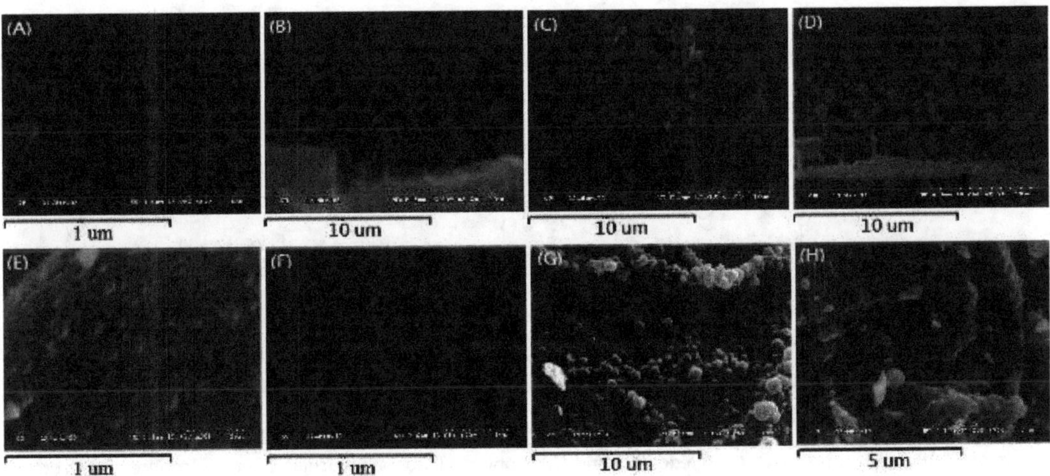

Figure 2. SEM image of GSCCMs of (A) SC500, (B) SC800, (C) SC1200, (D) SC500, (E) SE500, (F) SE800, (G) SE1200, and (H) SE1500.

corresponding to oxide layers was formed. The phenomenon can be attributed to oxidation in part of the carbon structure that converts to epoxy groups. The FT-IR spectra of our produced GSCCMs are shown in Figures 5 and 6. Compared with Figure 4, only peaks at 1700 to 1550 cm^{-1} and 878 to 590 cm^{-1} have been changed significantly with the changing of the heat-treatment temperatures.

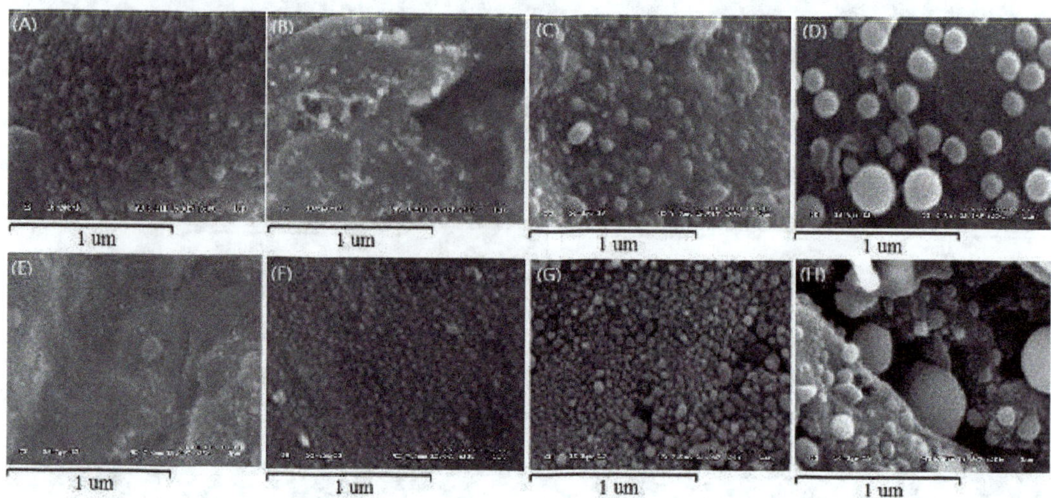

Figure 3. SEM image of GSCCMs of (A) AC500, (B) AC800, (C) AC1200, (D) AC500, (E) AE500, (F) AE800, (G) AE1200, and (H) AE1500.

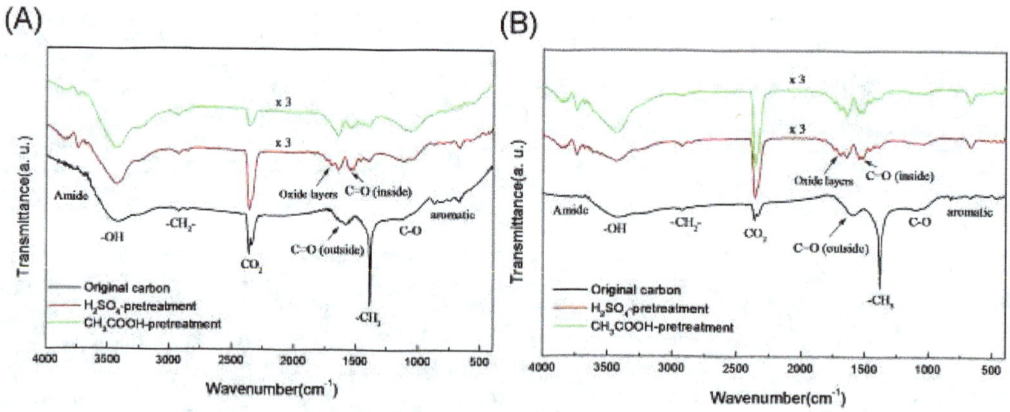

Figure 4. FT-IR spectrums of (A) dicotyledonous and (B) monocotyledonous biochars after acid pretreatment before heat treatment.

Figures 7–9 show the XRD spectrometry analysis results. The biochar and GSCCM samples have a peak of cellulose at around 2θ degree = $22°$, at around 2θ degree = $26°$ (d_{002}), $41°$ (d_{100}), and $44°$ (d_{101}) corresponding with crystals of graphite, and a peak at around 2θ degree = $31°$ corresponding with crystals of C_{60}. The peak (2θ degree = $22°$) has an obvious decrease in the intensity. This phenomenon is attribute to the disintegration of cellulose during the acid pretreatment process and the formation of graphitic crystals and oxide layers (2θ degree = $29°$) in the GSCCM samples.

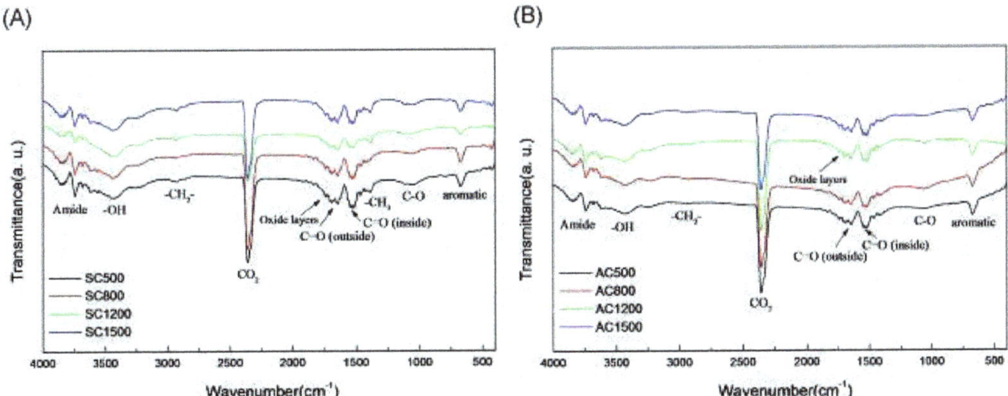

Figure 5. FT-IR spectrums of GSCCMs of the H_2SO_4 pretreated (A) dicotyledonous and (B) monocotyledonous biochars and heat-treated at various temperatures.

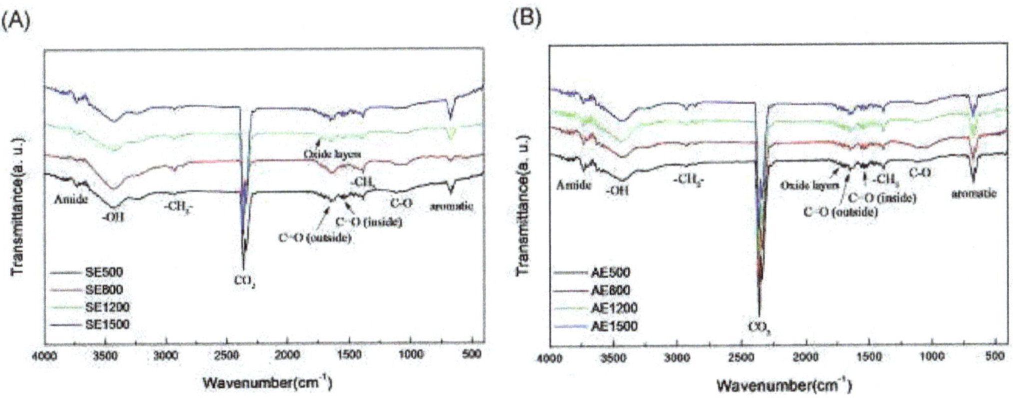

Figure 6. FT-IR spectrums of GSCCMs of the CH_3COOH pretreated (A) dicotyledonous and (B) monocotyledonous biochars after heat-treated at various temperatures.

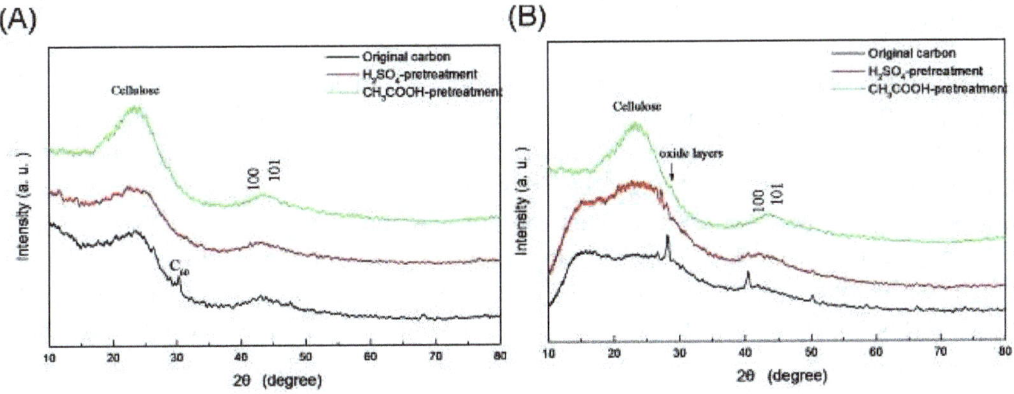

Figure 7. XRD spectrums of (A) dicotyledonous and (B) monocotyledonous biochars after acid pretreatment before heat treatment.

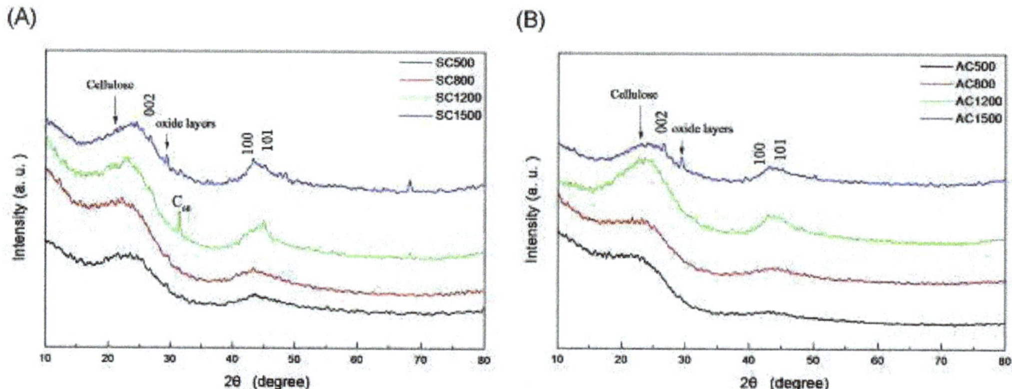

Figure 8. XRD spectrums of GSCCMs of the H_2SO_4 pretreated (A) dicotyledonous and (B) monocotyledonous biochars and heat-treated at various temperatures.

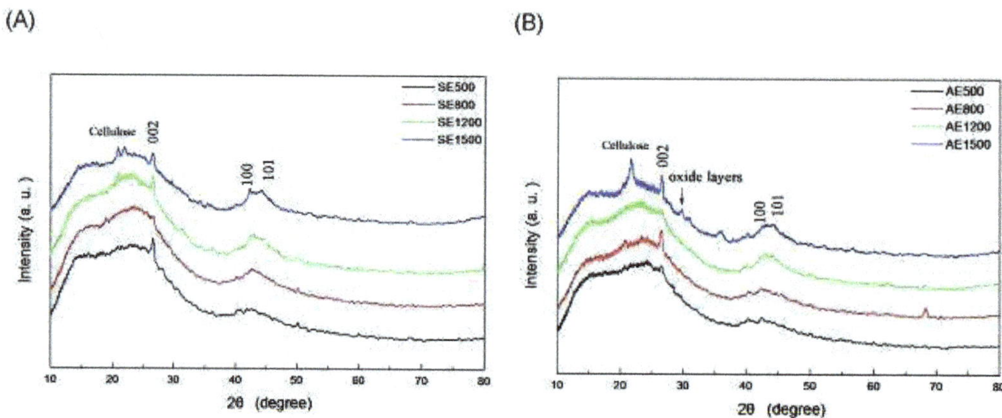

Figure 9. XRD GSCCMs of the CH_3COOH pretreated (A) dicotyledonous and (B) monocotyledonous biochars and heat-treated at various temperatures.

The graphene sheet content of produced GSCCM samples can be characterized quantitatively, according to the author's previous report [12]. Results are shown in Table 1, the GSCCMs, through the two-step processes from biochars, have an increment on the graphene sheet content of 16% to 28% more than that of the two-step heat-treatment-only process from woody biomass [15]. The highest graphene sheet content was found with CH_3COOH and H_2SO_4 pretreatment followed by 1500°C heat-treated monocotyledonous biochar materials; and its average conductivity was measured as 84.69 and 75.55 S/cm, respectively. The total metal content in our synthesized GSCCMs was almost less than 1 wt%, and its contribution to conductivity could be ignored (data not shown here).

According to the industry reports [16], the cost of artificial graphite and graphene for the CVD process is 1450 USD/ton and 28.57 USD/g, respectively. The production cost of GSCCMs was

Biomass	Dicotyledonous				Monocotyledonous			
Preparation method	A) Two stepwise heat treatment processes (without catalytic) [14] from woody biomass B) Two stepwise heat treatment processes (with catalytic) [14] from woody biomass C) Two-step process (H_2SO_4 pretreatment followed heat treatment) from biochar D) Two-step process (CH_3COOH pretreatment followed heat treatment) from biochar							
Method type	A	B	C	D	A	B	C	D
Graphene sheet content (%)								
500°C	18.65	24.66	22.29	25.14	10.08	28.80	22.34	22.56
800°C	23.62	26.08	25.69	25.09	25.00	32.00	36.87	30.61
1200°C	32.96	43.04	43.17	48.61	44.60	50.20	42.13	67.54
1500°C	56.06	62.08	63.12	72.18	80.60	81.20	79.71	83.86
Average conductivity (S/cm)								
500°C	0.01	0.17	0.12	0.14	0.03	0.27	0.28	0.01
800°C	1.24	0.98	0.19	0.13	0.06	0.81	0.93	0.52
1200°C	23.67	24.16	23.01	31.58	30.15	39.54	29.95	61.29
1500°C	50.01	53.31	59.02	66.45	79.08	82.40	75.55	84.69

Table 1. Electrical and graphene sheet content analysis of GSCCMs samples

about 13.33 TND/batch. Therefore, preparing GSCCMs from biochar materials could highly reduce the cost and could be applied in transparent conductive thin film, dry-sensitized solar cells, super capacitors, and composites.

4. Conclusions

According to our results, the highest graphene sheet content of 83.86% is found with the CH3COOH pretreatment followed by a 1500°C heat treatment of monocotyledonous biochar materials, and its conductivity was measured at 84.69 S/cm. Therefore, preparing GSCCMs from biochar materials could highly reduce the cost and could be applied in transparent conductive thin film, dry-sensitized solar cells, super capacitors, and composites.

Supplementary materials

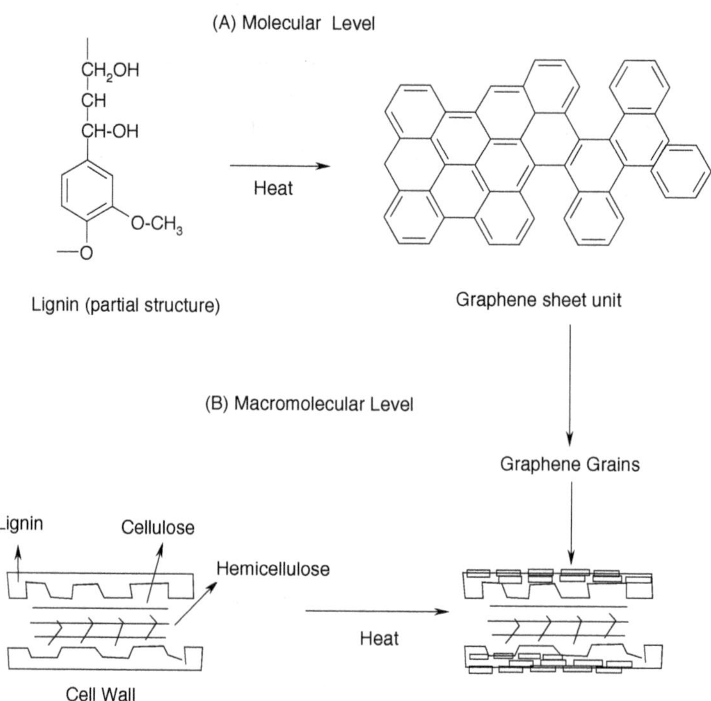

Figure S1. Proposed formation mechanism of graphene grains on cytoskeleton surface of wood cells. (Y. J. Liou, W. J. Huang. Relationship between the formation of graphene fraction and the development of surface electro-conductivity of pyrolized monocotyledonous and dicotyledonous plants. *J. Cryst. Growth,* Under review.)

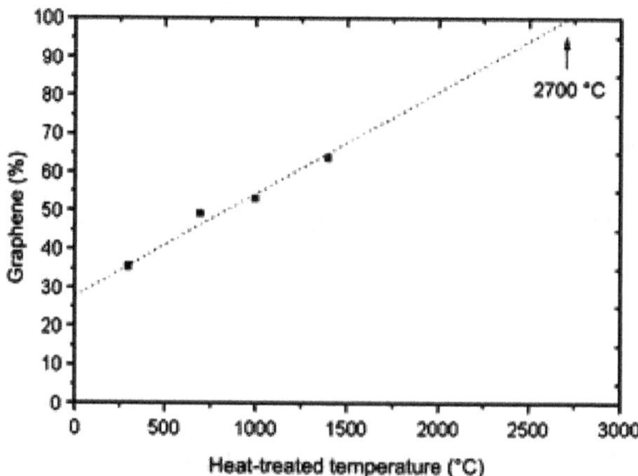

Figure S2. Plot of graphene content and heat-treated temperature of pyrolized wood char powders. (Y. J. Liou, W. J. Huang. Quantitative analysis of graphene sheet content in wood char powders during secondary graphitization. *J. Mater. Sci. Technol.,* 2013, 29: 406.)

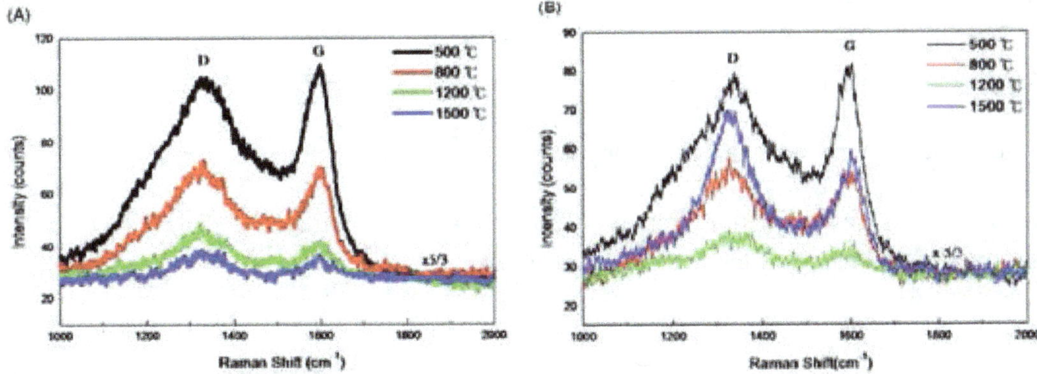

Figure S3. Raman spectrums of graphene sheet content carbon materials by (A) noncatalytic and (B) catalytic pyrolized processes. (Y. J. Liou, W. J. Huang. Determination of graphene sheets content in carbon materials by Raman spectroscopy. *J. Chin. Chem. Soc.* 2014, 61: 1045.)

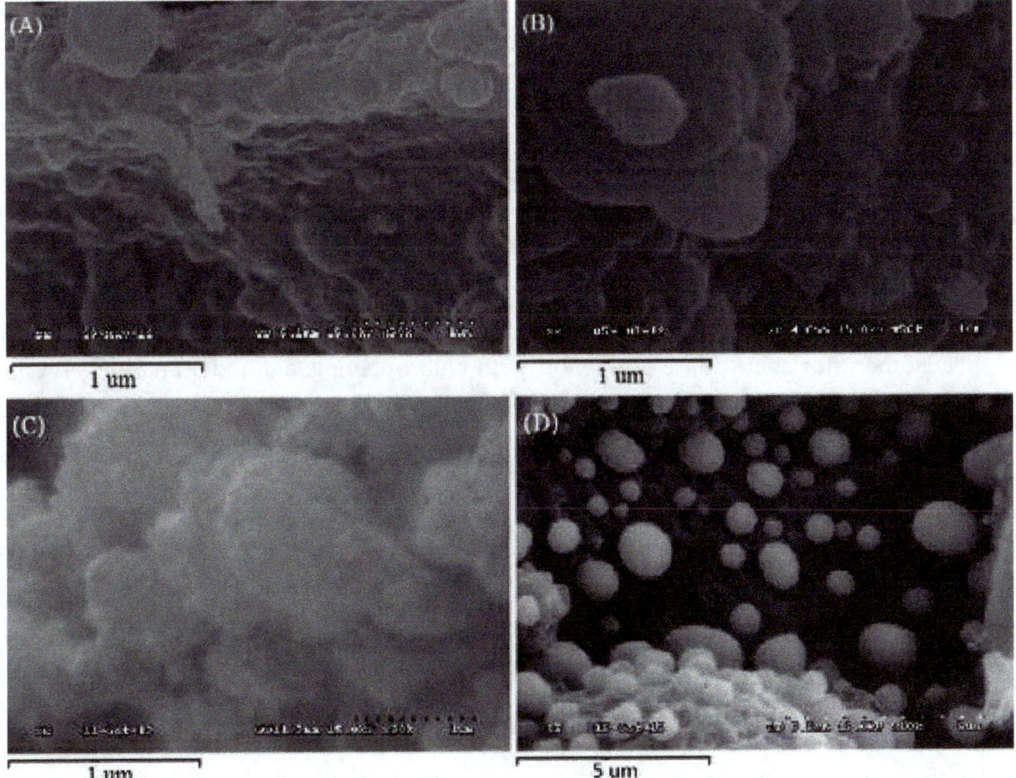

Figure S4. SEM image of GSCCMs by (A) noncatalytic, (B) MnO_2, (C) Li_2O, and (D) Na_2O catalytic pyrolized processes at 1500°C. (Yan-Jia Liou, Wu-Jang Huang. The role of sodium ions on the development of micro-sized pores in a high alpha-cellulose content woody biomass under ambient temperature. J. Cleaner Product. 2014, 47: 199.)

Author details

Yan-Jia Liou and Wu-Jang Huang*

*Address all correspondence to: wjhuang@mail.npust.edu.tw

Department of Environmental Engineering and Science, National Ping-Tung University of Science and Technology, Ping-Tung County, Taiwan

References

[1] Y. Lee, S. Ba, H. Jang, S. Jang, et al. Wafer-scale synthesis and transfer of graphene films. *Nano Lett.*, 2010, 10: 490.

[2] W. Hong, Y. Xu, G. Lu, G. Shi, et al. Transparent graphene/PEDOT-PSS composite films as counter electrodes of dye-sensitized solar cells. *Electrochem. Commun.*, 2008, 10: 1555.

[3] L. M. Hollanda, A. O. Lobo, M. Lancellotti, E. Berni, et al. Graphene and carbon nanotube nanocomposite for gene transfection. *Mater. Sci. Eng. C*, 2014, 39: 288.

[4] S. Sheshmani, A. Ashori, S. Hasanzadeh. Removal of acid orange 7 from aqueous solution using magnetic graphene/chitosan: a promising nano-adsorbent. *Int. J. Biol. Macromol.*, 2014, 68: 218.

[5] Q. Liu, L. Shi, L. Zeng, T. Wang, et al. Evaluation of graphene as an advantageous adsorbent for solid-phase extraction with chlorophenols as model analytes. *J. Chromatogr. A*, 2011, 1218: 197.

[6] J. J. Lu, B. W. Qiu, K. P. Huang, Z. Z. Chang. Large-area synthesis of grapheme monolayer by chemical vapor deposition. *J. Mac. Ind.*, 2010, 326: 103.

[7] X. Shen, L. Jiang, Z. Ji, J. Wu, et al. Stable aqueous dispersions of graphene prepared with hexamethylenetetramine as a reductant. *J. Coll. and Int. Sci.* 2011, 354: 493.

[8] K. S. Novoselov, A. K. Geim, S. V. Morozov, D. Jiang, et al. Electric field effect in atomically thin carbon films. *Science.* 2004, 306(5695): 666.

[9] A. K. Geim. Graphene: status and prospects. *Science.* 2009, 324: 1530.

[10] M. J. Webb, P. Palmgren, P. Pal, O. Karis, et al. A simple method to produce almost perfect graphene on highly oriented pyrolytic graphite. *Carbon*, 2011, 49: 3242.

[11] O. Akhavan. The effect of heat treatment on formation of graphene thin films from graphene oxide nanosheets. *Carbon*, 2010, 48: 509.

[12] Y. J. Liou, W. J. Huang. Quantitative analysis of graphene sheet content in wood char powders during secondary graphitization. *J. Mater. Sci. Technol.*, 2013, 29: 406.

[13] Y. J. Liou, W. J. Huang. Determination of graphene sheets content in carbon materials by Raman spectroscopy. *J. Chin. Chem. Soc.* 2014, 61: 1045.

[14] Yan-Jia Liou, Wu-Jang Huang. The role of sodium ions on the development of micro-sized pores in a high alpha-cellulose content woody biomass under ambient temperature. *J. Cleaner Product.* 2014, 47: 199.

[15] Y. J. Liou, W. J. Huang. Relationship between the formation of graphene fraction and the development of surface electro-conductivity of pyrolized monocotyledonous and dicotyledonous plants. *J. Cryst. Growth.* Under review.

[16] http://big5.mofcom.gov.cn/gate/big5/price.mofcom.gov.cn/channel/priceinfo.shtml?prod_id=0111651.

8

Nucleation Behaviors of Nd and Dy in TFSA-Based Ionic Liquids

Masahiko Matsumiya

Abstract

The nucleation behavior of $[Nd^{(III)}(TFSA)_5]^{2-}$ and $[Dy^{(III)}(TFSA)_5]^{2-}$ in the TFSA-based ionic liquid (IL), triethyl-pentyl-phosphonium bis(trifluoromethyl-sulfonyl)amide, $[P_{2225}][TFSA]$, was investigated in this study. The initial process of Nd and Dy electrodeposition was evaluated by chronoamperometry, indicating that the initial nucleation and the growth of Nd and Dy on the electrode surface occurred via instantaneous nucleation at -3.40 and -3.60 V, respectively. As the overpotential induced more negative, the nucleation mechanism altered from instantaneous to progressive. The number density of Nd and Dy nuclei tended to decrease as the overpotential gradually increased in this system. Moreover, the potentiostatic electrodeposition of Nd and Dy metals was examined at 393 K. The surface morphology of the electrodeposits was consistent with the chronoamperometric results. From the EDX and the XPS analyses, we ascertained that the main electrodeposits were rare earth metals with a small quantity of light elements. The series of results enabled us to conclude that the greater part of the electrodeposited Nd and Dy metals was obtained from TFSA-based IL bath by potentiostatic electrodeposition with elevating temperatures, and the control of the water content of the electrolyte was an important factor for the recovery of metallic Nd and Dy with high purity.

Keywords: Electrodeposition, ionic liquids, nucleation, rare earths

1. Introduction

Rare earth (RE) elements have been practically applied in various high-tech industries as magnetic and electronic materials [1-3] owing to their unique characteristics deriving from the 4f electron structure. Neodymium metal in particular is an especially important element as it is the main raw material for permanent Nd-Fe-B magnets used in hard disk drives, mobile phones, and automotive motors [4-7]. A large number of Nd-Fe-B magnets have been produced in the world in recent years, and a significant number of spent Nd-Fe-B magnets will accumulate in the near future. It is therefore necessary to recover the RE metals from these secondary waste magnets. Electrowinning methods with high-temperature molten salts have been employed as a conventional recovery process for RE metals because the reduction potential of RE metals is extremely negative. The high-temperature systems needed for these methods for the last several decades. Although the pyrometallurgical process [8-10] of RE metals has actually developed so far, it is difficult to establish a practical recovery system. This kind of pyrometallurgical process using high-temperature media consumed extraordinary amounts of thermal energy, so that it would not be appropriate as an environmental-friendly process in the near future. Therefore, the development of an environmentally conscious technology for the recovery of RE metals is desirable.

From the standpoint of environmental harmonization, the electrodeposition using ionic liquids (ILs) is attractive as the effective recovery method of Nd and Dy metals. ILs have many original properties, such as nonvolatility, noncombustible, and a wide electrochemical window [11, 12]. In particular, the ILs consisting of bis(trifluoromethyl-sulfonyl)amide anion, [TFSA$^-$], are easily dehydrated compared with other ILs that are miscible with water. There are some reports about the electrochemical behavior and electrodeposition of RE elements such as La, Sm, Eu, and Yb [13-17] in ILs. We have already demonstrated that the recovery process of Nd [18-21] and Dy [22, 23] metals by electrodeposition using quaternary phosphonium and ammonium-based ILs, such as triethyl-pentyl-phosphonium bis(trifluoromethyl-sulfonyl)amide, [P$_{2225}$][TFSA], and 2-hydroxy ethyl-trimethyl-ammonium bis(trifluoromethyl-sulfonyl)amide, [N$_{1112OH}$][TFSA]. In addition to the electrodeposition process using ILs, the wet separation process has been recently indispensable for the recycling of spent Nd-Fe-B magnets. In particular, solvent extraction and precipitation separation are effective wet separation processes. Therefore, we have recently proposed the combination process constituting a wet separation, such as solvent extraction [24, 25] or precipitation separation [26, 27], with electrodeposition [18-23].

In this study, we investigated the nucleation behavior for the initial electrodeposition for Nd and Dy metals in TFSA-based ILs. In addition, the nucleation and the growth mechanisms of Nd and Dy nuclei were analyzed by chronoamperometry (CA). Furthermore, the potentiostatic electrodeposition of Nd and Dy metals from TFSA-based ILs was performed, and the obtained electrodeposits were characterized via scanning electron microscopy (SEM), energy-dispersive X-ray microanalysis (EDX), and X-ray photoelectron spectroscopy (XPS).

2. Experimental

2.1. Preparation

Nd(TFSA)$_3$ and Dy(TFSA)$_3$ were prepared from the reaction for Nd$_2$O$_3$ (Wako Pure Chemical Industries, Ltd., >99.9%) or Dy$_2$O$_3$ (Wako Pure Chemical Industries, Ltd., >99.9%) with 1, 1, 1-trifluoro-N-[(trifluoromethyl)sulfonyl]methanesulfone-amide (H[TFSA], Kanto Chemical Co.,

Inc., >99.0%) at 373 K under agitation. The suspension was clearly transformed to a pale purple (Nd) or a light-yellow (Dy) solution after the reaction between the metallic Nd and Dy components and the amide acid. Each solution was evaporated at 423 K in order to remove the unreacted acid components. $Nd(TFSA)_3$ and $Dy(TFSA)_3$ were obtained as a purplish and a yellowish fine powders and dried at 373 K in a vacuum chamber (<-0.1 MPa) for more than 48 h; the average yield of the synthesized $Nd(TFSA)_3$ and $Dy(TFSA)_3$ was confirmed to be >92.0%.

The phosphonium type of IL, $[P_{2225}][TFSA]$, was synthesized based on the previous procedure described by Tsunashima [28]. The metathesis reaction between $[P_{2225}]Br$ (Nippon Chemical Industrial Co., Ltd., >99.5%) and $Li[TFSA]$ (Kanto Chemical Co., Inc., 99.7%) proceeded to obtain $[P_{2225}][TFSA]$. The crude IL was purified by cleaning several times with distilled water. All bromide anions were perfectly removed from the purified IL, as confirmed by $AgNO_3$ titration. $[P_{2225}][TFSA]$ prepared from this synthesis was dried under high vacuum (<–0.1 MPa) for 72 h at 393 K. The average yield of IL based on the synthesis was 96.9%. A solution of $Nd(TFSA)_3$ and $Dy(TFSA)_3$ in $[P_{2225}][TFSA]$ was used as the electrolyte for the ensuing experiment and was prepared by adding an appropriate amount of $Nd(TFSA)_3$ or $Dy(TFSA)_3$ to $[P_{2225}][TFSA]$ inside a glove box (MIWA MFG Co., Ltd., DBO-2LKP-YUM02) filled with Ar (water and oxygen contents <1.0 ppm). These solutions were dried at 373 K in a vacuum chamber (<-0.1 MPa) for more than 72 h.

2.2. Electrochemical analysis

CA was carried out using a three-electrode cell with an electrochemical analyzer (BAS Inc., ALS760D) inside a glove box. The Pt disk electrode (BAS Inc., ϕ = 1.6 mm) was selected as a working electrode that was mirror polished using alumina and diamond pastes. Pt wire (BAS Inc., ϕ = 0.5 mm) was used as a counter electrode for CA. Ag wire immersed in 0.1 mol dm^{-3} $Ag[CF_3SO_3]$ (Wako Pure Chemical Industries, Ltd., >97.0%) in $[P_{2225}][TFSA]$ was applied as a reference electrode. The potential was compensated for the IL standard using a ferrocene (Fc)/ ferrocenium (Fc$^+$) redox couple.

The electrodeposition of Nd and Dy metals from 0.1 mol dm^{-3} $Nd(TFSA)_3$ and $Dy(TFSA)_3$ in $[P_{2225}][TFSA]$ was carried out using a three-electrode system inside a cylindrical cell at 393 K under an Ar atmosphere (H_2O, O_2 < 1.0 ppm) in a glove box. A copper substrate and a prismatic Nd-Fe-B rod were employed as a cathode and an anode, respectively. A Vycor glass filter was covered with a soda lime tube at the bottom of anode in order to prevent the diffusion of dissolution components from the anode into the electrolyte. The electrode surface was polished by the suitable fine-grade abrasive papers before experiments. The Pt wire (ϕ = 0.5 mm) was used as a quasi-reference electrode (QRE). Moreover, the overpotential under the potentio-static electrodeposition was -3.40 V or -3.80 V for the electrodeposition of Nd or Dy metal, respectively. The surface morphology of the electrodeposits was observed using SEM (JSM-6510LA, JEOL Ltd.), and the chemical composition of electrodeposits was analyzed using EDX (JED-2300, JEOL Ltd.). The metallic states of Nd and Dy element in the electrodeposits were evaluated using XPS (Quantera SXM, ULVAC-PHI, Inc.). Furthermore, the crystalline state of the metallic Nd and Dy was investigated by XRD (RINT-2500, Rigaku Co.).

3. Results and discussions

3.1. Nucleation behaviors of Nd and Dy nuclei

According to the chronoamperometric analysis, the initial nucleation and the subsequent growth behaviors of Nd and Dy nuclei were investigated in $[P_{2225}][TFSA]$. A distinctive peak was observed on the nucleation process after the current extinction corresponding to the double layer charging. This peak was characteristic of the three-dimensional growth of the Nd and Dy nuclei; there was an increase tendency of faradaic current based on the nucleation and the subsequent growth of Nd and Dy nuclei. Eventually, a maximum current density, j_m, was reached at a maximum time t_m; finally, the current immediately decreased to a limiting diffusion current. As for Nd nucleation, chronoamperogram was represented with different overpotentials, as shown in Fig. 1.

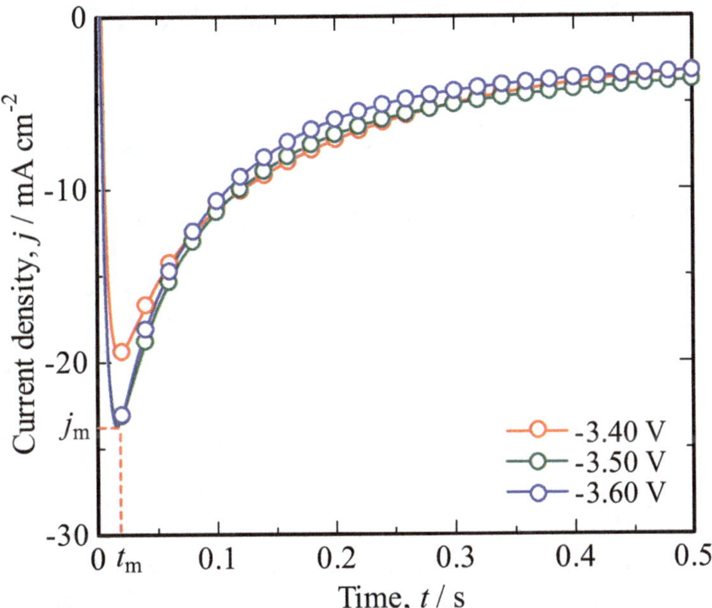

Figure 1. Chronoamperogram on a Pt electrode for 0.1 mol dm^{-3} [Nd$^{(III)}$(TFSA)$_5$]$^{2-}$ in [P$_{2225}$][TFSA] at 353 K.

For the chronoamperogram, the charging current of neat $[P_{2225}][TFSA]$ was deduced from the raw data for $[P_{2225}][TFSA]$ containing 0.1 mol dm^{-3} [Nd$^{(III)}$(TFSA)$_5$]$^{2-}$ in order to analyze the faradaic current only in the reduction reaction of [Nd$^{(III)}$(TFSA)$_5$]$^{2-}$. Several models have been developed to describe the j-t curves for metal deposition. It has been previously reported that the theoretical methodology developed by Scharifker and Hills [29] is applicable to deposition of several metals from TFSA-based ILs [30-34]. Therefore, the initial stage of nucleation and the crystal growth of Nd and Dy nuclei were evaluated based on the Hills-Scharifker theory in this investigation. The previous theory related with the three-dimensional nucleation can

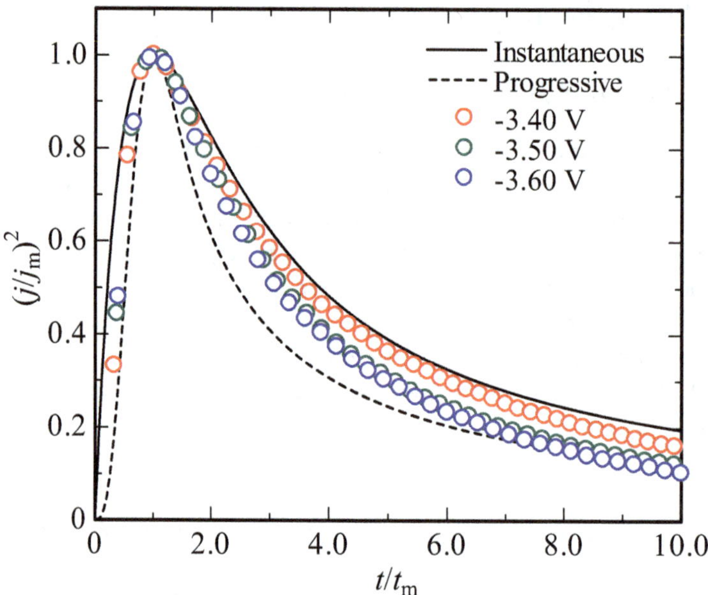

Figure 2. $(j/j_m)^2$-t/t_m plots for 0.1 mol dm^{-3} [Nd$^{(III)}$(TFSA)$_5$]$^{2-}$ in [P$_{2225}$][TFSA] evaluated from CA.

be classified into instantaneous and progressive process. As for the instantaneous nucleation, the nuclei are electrodeposited on the electrode active surface, and the nucleation growth proceeds at a constant rate that is dependent on the induced overpotential. The resultant j-t curves are mathematically described by the following equation:

$$\left(\frac{j}{j_m}\right)^2 = \frac{1.2254}{t/t_m}\left\{1-\exp\left[-2.3367\left(\frac{t}{t_m}\right)^2\right]\right\}^2 \tag{1}$$

where j is the current density at time t and j_m is the maximum current density at time t_m. In the case of progressive nucleation, the initially deposited nuclei grow at varying rates dependent on the time of deposition. The j-t curves can be mathematically described by the following equation:

$$\left(\frac{j}{j_m}\right)^2 = \frac{1.9542}{t/t_m}\left\{1-\exp\left[-1.2564\left(\frac{t}{t_m}\right)\right]\right\}^2 \tag{2}$$

The relationship between the squared dimensionless current density and the dimensionless time of [Nd$^{(III)}$(TFSA)$_5$]$^{2-}$ in [P$_{2225}$][TFSA] is shown in Fig. 2. In the case of the overpotential at -3.40 V, the current tendency was consistent with the instantaneous nucleation process.

Conversely, when the overpotentials were induced at -3.50 and -3.60 V, the current curves shifted from that of the instantaneous nucleation process to the progressive nucleation process. A series of experimental results demonstrated that the nucleation process influenced on the induced overpotentials. Thus, the nucleation and the growth process would be correlated with the number density of the active sites from the current transients [35] as discussed below. This transition from instantaneous to progressive nucleation was consistent with the nucleation process for Dy nuclei in our investigations [22]. In the case of Dy nuclei in $[P_{2225}][TFSA]$ system, when the overpotential of -3.60 V was induced, the nucleation behavior obeyed the instantaneous nucleation process. On the other hand, the nucleation curve at -3.80 V was followed by the progressive nucleation process.

The number density was analyzed based on the following procedure: the nucleation number density, N_0, can be estimated from the maximum current, j_{max}, and the maximum time, t_{max}, of the j-t curve based on the following equation [35]:

$$N_0 = 0.065\left(8\pi C_O^* M/\rho\right)^{-1/2}\left(nFC_O^*/j_{max}t_{max}\right)^2 \tag{3}$$

where M and ρ are the atomic mass and the density of the deposited metal, respectively. During the initial stages of electrodeposition, well before the maxima in the j-t curves, the initial nucleation can be regarded as an effectively instantaneous, i.e., all related nuclei would be considered to have been formed simultaneously [36]. Then, Eq. (3) can be applied to evaluate the value of N_0 for all systems from the early stage of the current transient. The number of Nd and Dy nuclei formed in the early stages of nucleation was estimated using this method as listed in Table 1. In general, the nucleation number density would be tended to increase with elevating overpotential on progressive nucleation process, e.g., the Sn nucleation [37] in 1-ethyl-3-methylimidazolium chloride, tetrafluoro-borate, $[C_2C_1Im][Cl, BF_4]$. Contrary to our expectations, the number density of Nd and Dy nuclei tended to decrease as the overpotential gradually induced in this study. This phenomenon would be correlated with the transition from instantaneous to progressive process. It was also reported that the rate of nucleation was influenced from the overpotential for instantaneous or progressive process [38]. Thus, the different nucleation rate on instantaneous or progressive process would be related with the nucleation number density. The mechanism of Nd and Dy [22] nucleation revealed that the number density of nuclei is dependent on the overpotential, suggesting that the applied overpotential for electrodeposition becomes important factor from the perspective of the nucleation mechanism as well as the number of nuclei.

Applied potential (E) for Nd nucleation/V	-3.40	-3.50	-3.60
Nucleation number density ($10^{11} N_0$)/cm⁻²	2.0	1.7	1.7
Applied potential (E) for Dy nucleation/V	-3.40	-3.60	-3.80
Nucleation number density ($10^{11} N_0$)/cm⁻²	6.3	5.5	2.6

Table 1. The nucleation number density of Nd and Dy nuclei from j-t curves at each applied potential.

3.2. Electrodeposition of Nd and Dy metals

Electrodeposition at -3.40 V produced blackish electrodeposits on a Cu substrate. The total transported charge was 105.8 C. The SEM image of the electrodeposits was shown in Fig. 3.

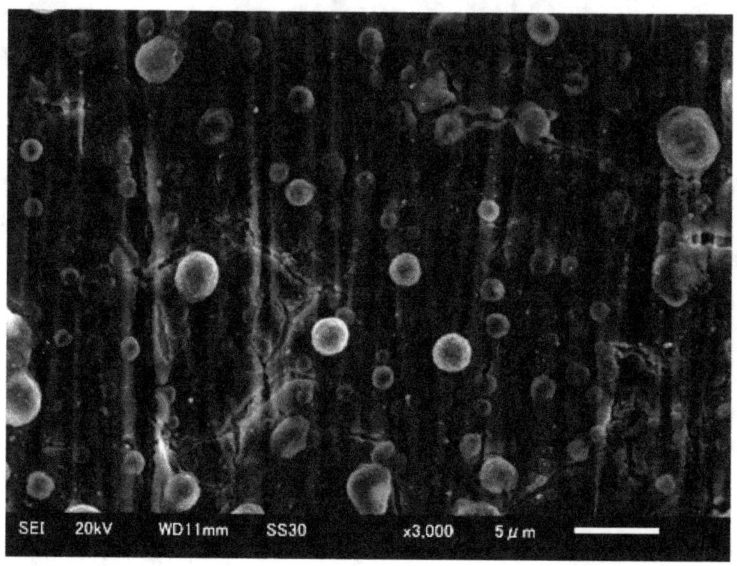

Figure 3. The SEM image of electrodeposits obtained by potentiostatic electrodeposition induced at -3.40 V.

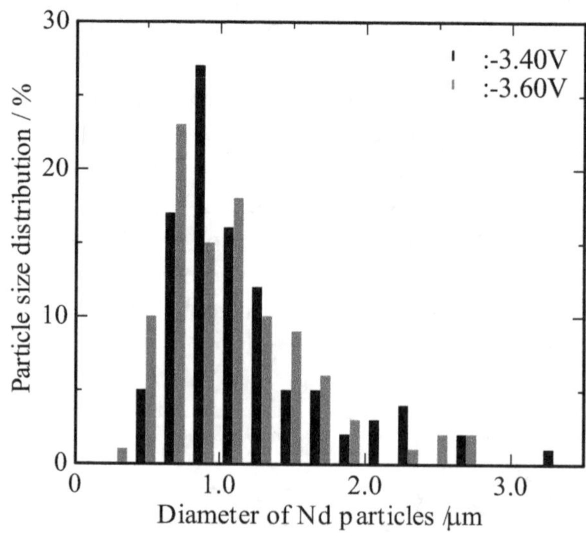

Figure 4. The size distribution of the electrodeposited Nd particles evaluated from each SEM image.

The surface morphology of the electrodeposits was heterogeneous orbicular with a nonuniform size distribution. This behavior would be generated from the initial nucleation and the subsequent growth occurred according to the progressive nucleation process in which the rate of the nucleation was faster than that of the crystal growth. From above-mentioned nucleation behaviors, at -3.40 V, the j-t curve would be fitted to the theoretical line for instantaneous nucleation process, but the line gradually shifted to the progressive nucleation process. As a result, the electrodeposition result was consistent with the data for the initial nucleation and growth of Nd nuclei in [P$_{2225}$][TFSA] by CA discussed above. With the use of applied overpotentials of -3.40 V, the average diameter of the Nd particles was 1.17 μm, as shown in Fig.4. The particle size of the electrodeposits obtained by electrodeposition at -3.60 V was smaller than that achieved at -3.40 V. The number of particles electrodeposited at -3.60 V was also larger than that at -3.40 V. A series of results were also consistent with the CA results, which represented that the number density of Nd nuclei tended to increase with elevating the overpotential. The quantitative analysis for the electrodeposits at -3.40 and -3.60 V evaluated from EDX was listed in Table 2.

Overpotential/V	C	N	O	F	S	Fe	Nd
-3.40	1.26	0.94	20.36	2.83	0.89	0.23	73.49
-3.60	1.15	0.78	23.69	3.26	1.21	0.38	69.53

Table 2. The composition (wt.%) of the electrodeposits obtained by potentiostatic electrodeposition induced at -3.40 and -3.60 V from 0.1 mol dm^{-3} [Nd$^{(III)}$(TFSA)$_5$]^2 in [P$_{2225}$][TFSA] at 353 K.

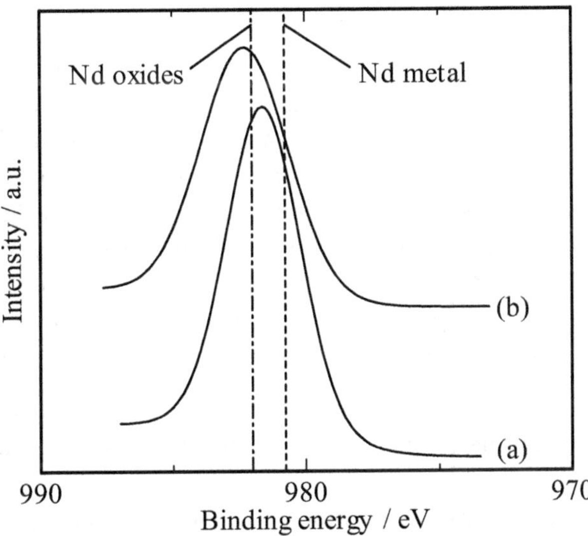

Figure 5. The Nd 3d$_{5/2}$ spectra of the electrodeposits obtained by XPS analysis. (a) The spectrum for middle layer under 0.75 μm of the electrodeposits. (b) The spectrum for the top surface of the electrodeposits.

The EDX data display electrodeposits at -3.40 and -3.60 V, which mainly consist of Nd metal. However, a relatively large amount of oxygen was also detected at the top surface in the electrodeposits, indicating that a part of Nd metal would be oxidized due to the active metal. Thus, the oxidizing state of Nd metal in the electrodeposits was evaluated from XPS with Ar etching. Fe element was scarcely detected in the electrodeposits at both overpotentials. As a result, the separated anode structure from the main electrolyte by the Vycor glass filter was confirmed to be an effective strategy. The decomposition of the ILs was prevented by this separated anode structure. Thus, this separated anode structure and the dissolution of anode compartment enabled the acquisition of Nd metal in the electrodeposits while preventing bonding with the light elements. The Nd $3d_{5/2}$ XPS spectra of the electrodeposits at -3.40 and -3.60 V were shown in Figs. 5(a) and 5(b), respectively. The electrodeposited surfaces were sputtered by an Ar ion beam to facilitate in-depth analysis of the middle layer, that is, (a) under 0.75 μm and (b) the top surface of the electrodeposits. The binding energy of Nd $3d_{5/2}$ for Nd metal and oxides should be theoretically positioned at 980.5-981.0 and 981.7-982.3 eV, respectively [39]. The Nd $3d_{5/2}$ peak maxima acquired for the middle layers (a) under 0.75 μm and (b) the top surface were detected at 981.57 and 982.25 eV, respectively. Hence, the electrodeposits were identified as Nd metal and oxide mixtures. It is conjectured that metallic Nd should initially be electrodeposited on a Cu substrate and subsequently oxidized by the oxygen in ILs. It has reported that the water content of the ILs did not significantly affect the oxidation sate of the electrodeposits but would influence the coordination number of the Nd complex [18]. Therefore, the sufficient dehydration and control of the water content of the ILs, including metallic components, are important factors for achieving the recovery of metallic Nd with high purity.

As for the electrodeposition of Dy metal, the potentiostatic electrodeposition at -3.80 V in $[P_{2225}][TFSA]$, including 0.1 mol dm^{-3} $[Dy^{(III)}(TFSA)_5]^{2-}$, was performed at 393 K, according to the above-mentioned nucleation analyses. The potentiostatic electrolysis was conducted in a glove box with stirring at 500 rpm to increase the current density by intermittently supplying the electroactive species $[Dy^{(III)}(TFSA)_5]^{2-}$ to the reactive electrode surface. The total transported charge under Dy electrodeposition was 300 C for 43.6 h. There was no decomposition reaction of IL under Dy electrodeposition since there was no alternation on the light-yellow electrolyte before and after electrodeposition. The blackish-brown electrodeposits with a strong adhesion were observed on a Cu cathode substrate. A microanalytical SEM/EDX method was applied to the analysis of the surface morphology and the element content of the electrodeposits. The EDX results with SEM image are shown in Figs. 6(a) and 6(b), respectively.

From Fig. 6(b), the round particles on a Cu substrate were confirmed to be Dy metal and other related compounds. A quantitative analysis of the electrodeposits is tabulated in Table 3.

Most parts of the electrodeposits were composed of the Dy metallic state because almost no oxygen was detected on the Dy metals from the EDX analysis. The amounts of carbon and oxygen were derived from the surface of the Cu cathode substrate. The Dy metals with ca. 0.8-0.9 μm particles were highly distributed in the electrodeposits. This electrodeposition

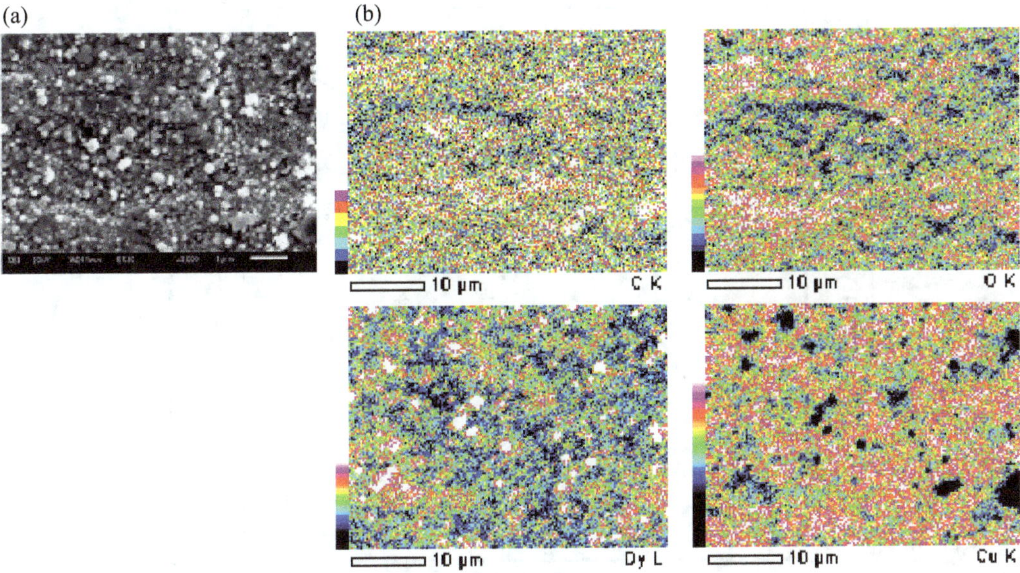

Figure 6. (a) The SEM image and (b) EDX analysis of Dy electrodeposits on a Cu substrate under potentiostatic electrodeposition at -3.80 V in $[P_{2225}][TFSA]$, including 0.1 mol dm^{-3} $[Dy^{(III)}(TFSA)_5]^{2-}$ at 393 K.

Overpotential/V	C	N	O	F	P	S	Dy	Cu
-3.80	10.83	1.15	11.82	3.87	0.86	1.87	9.27	60.33

Table 3. The composition (wt.%) of the electrodeposits on a Cu substrate obtained by potentiostatic electrodeposition induced at -3.80 V in $[P_{2225}][TFSA]$, including 0.1 mol dm^{-3} $[Dy^{(III)}(TFSA)_5]^{2-}$ at 393 K.

enabled us to conclude that Dy electrodeposits contained a smaller amount of the oxide state and a large amount of the metallic state on the electrode top surface. We also investigated the depth components of the electrodeposits by XPS with the Al-Kα radiation. The Dy $3d_{5/2}$ spectra for the top surface and the middle layer at the depth of 0.5 μm of the Dy electrodeposits are represented in Figs. 7(a) and 7(b), respectively. The XPS analysis with Ar sputtering at the rate of 27.2 nm min^{-1} was performed on the Dy electrodeposits. The Dy $3d_{5/2}$ spectra for Dy metal and oxides were theoretically positioned at 1295.8 and 1289.0 eV, respectively [30]. The Dy $3d_{5/2}$ peaks of the electrodeposits were assigned at 1297.2 eV and were hardly shifted by Ar etching treatment. The Dy $3d_{5/2}$ spectra of the middle layer and the top surface showed a good agreement with the theoretical value. The XRD result of the Dy electrodeposits is shown in Fig. 8. The Dy metallic crystal planes at a low angle were in good accordance with the ASTM card. Consequently, a series of electrodeposition investigations of Nd and Dy electrodeposits allowed us to conclude that most of the metallic state of Nd and Dy can be recovered from a TFSA-based IL by the potentiostatic electrodeposition with elevating temperatures.

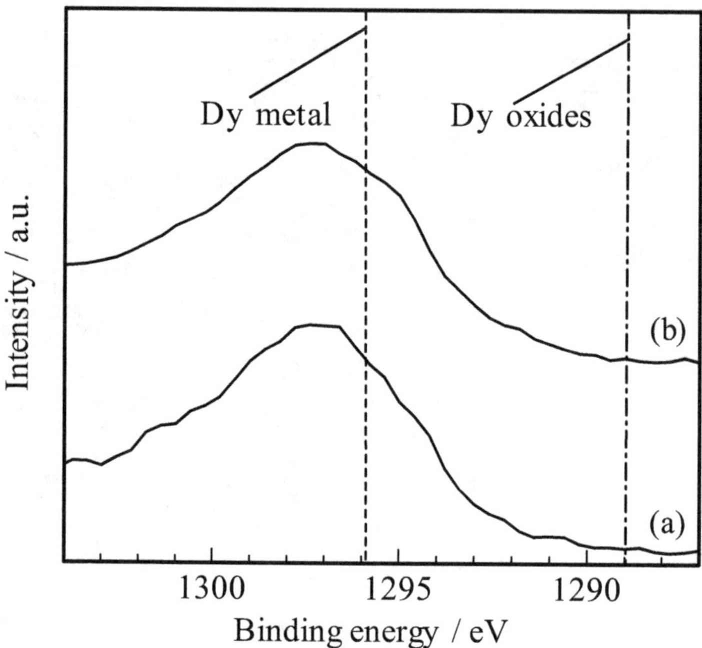

Figure 7. The Dy $3d_{5/2}$ spectra of the Dy electrodeposits by XPS analysis. (a) The top surface of the electrodeposits. (b) The middle layer under 0.5 μm of the electrodeposits.

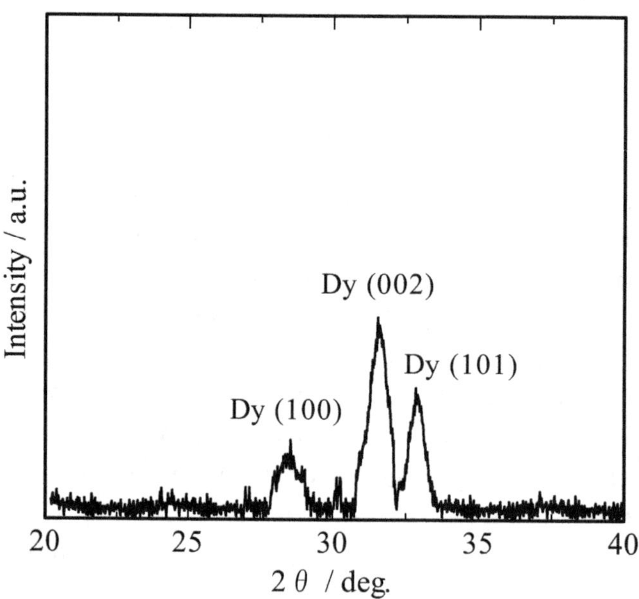

Figure 8. XRD profile of the Dy electrodeposits on a Cu substrate under potentiostatic electrodeposition at -3.80 V.

Acknowledgements

This study was partially supported by the Grant-in-Aid for Scientific Research (No. 15H02848) from the Ministry of Education, Culture, Sports, Science and Technology, Japan.

Author details

Masahiko Matsumiya

Address all correspondence to: mmatsumi@ynu.ac.jp

Graduate School of Environment and Information Sciences, Yokohama National University, Tokiwadai, Hodogaya-ku, Yokohama, Japan

References

[1] D. Yuan, Y. Liu, Mater. Chem. Phys., 96 (2006) 79-83.

[2] H. Yoshida, S. Kato, K. Hirano, J. Nishimoto, T. Hattori, Chem. Lett., 36(3) (2007) 430-431.

[3] A.A. Dakhel, Mater. Chem. Phys., 102 (2007) 266-270.

[4] N. Oono, M. Sagawa, R. Kasada, H. Matsui, A. Kimura, J. Magn. Magn. Mater., 323 (2011) 297-300.

[5] Y. Matsuura, J. Magn. Magn. Mater., 303 (2006) 344-347.

[6] O. Takeda, T. Okabe, Y. Umetsu, J. Alloys Compd., 408-412 (2006) 387-390.

[7] G. Yan, P.J. McGuiness, J.P.G. Farr, I.R. Harris, J. Alloys Compd., 478 (2009) 188-192.

[8] S. Kobayashi, K. Kobayashi, T. Nohira, R. Hagiwara, T. Oishi, H. Konishi, J. Electro-chem. Soc., 158 (2011) E142-E146.

[9] T. Uda, Mater. Trans., 43(1) (2002) 55-62.

[10] O. Takeda, T.H. Okabe, Y. Umetsu, J. Alloys Compd., 408-412 (2006) 387-390.

[11] D.R. MacFarlane, M. Forsyth, P.C. Howlett, J.M. Pringle, J. Sun, G. Annat, W. Neil, E.I. Izgorodina, Acc. Chem. Res., 40 (2007) 1165-1173.

[12] W. Simka, D. Puszczyk, G. Nawrat, Electrochim. Acta, 54 (2009) 5307-5319.

[13] S. Legeai, S. Diliberto, N. Stein, C. Boulanger, J. Estager, N. Papaiconomou, M. Draye, Electrochem. Commun., 10 (2008) 1661-1664.

[14] M. Yamagata, Y. Katayama, Y. Miura, J. Electrochem. Soc., 153 (2006) E5-E9.

[15] A. I. Bhatt, I. May, V. A. Volkovich, D. Collison, M. Helliwell, I. B. Polovov, R. G. Lewin, Inorg. Chem., 44(14) (2005) 4934-4940.

[16] K. Binne, Chem. Rev., 107 (2007) 2592-2614.

[17] Y. Katayama, Electrochemical Aspects of Ionic Liquids, (2005) 111-131.

[18] H. Kondo, M. Matsumiya, K. Tsunashima, S. Kodama, Electrochim. Acta, 66 (2012) 313-319.

[19] H. Kondo, M. Matsumiya, K. Tsunashima, S. Kodama, ECS Trans., 50(11) (2012) 529-538.

[20] M. Ishii, M. Matsumiya, S. Kawakami, ECS Trans., 50(11) (2012) 549-560.

[21] M. Matsumiya, M. Ishii, R. Kazama, S. Kawakami, Electrochim. Acta, 146 (2014) 371-377.

[22] R. Kazama, M. Matsumiya, N. Tsuda, K. Tsunashima, Electrochim. Acta, 113 (2013) 269-279.

[23] A. Kurachi, M. Matsumiya, K. Tsunashima, S. Kodama, J. Appl. Electrochem., 42 (2012) 961-968.

[24] M. Matsumiya, Y. Kikuchi, T. Yamada, S. Kawakami, Sep. Purif. Technol., 130 (2014) 91-101.

[25] Y. Kikuchi, M. Matsumiya, S. Kawakami, Solv. Extr. Res. Dev., 21(2) (2014) 137-145.

[26] K. Ishioka, M. Matsumiya, M. Ishii, S. Kawakami, Hydrometallurgy, 144-145 (2014) 186-194.

[27] M. Matsumiya, K. Ishioka, T. Yamada, M. Ishii, S. Kawakami, Inter. J. Miner. Process., 126 (2014) 62-69.

[28] K. Tsunashima, M. Sugiya, Electrochem. Commun., 9 (2007) 2353-2358.

[29] B. Scharifker, G. Hills, Electrochim. Acta, 28 (1983) 879-889.

[30] Y. Bando, Y. Katayama, T. Miura, Electrochim. Acta, 53 (2007) 87-91.

[31] N. Tachikawa, N. Serizawa, Y. Katayama, T. Miura, Electrochim. Acta, 53 (2008) 6530-6534.

[32] Y.-L. Zhu, Y. Kozuma, Y. Katayama, T. Miura, Electrochim. Acta, 54 (2009) 7502-7506.

[33] A. Basile, A.I. Bhatt, A.P. O'Mullane, S.K. Bhargava, Electrochim. Acta, 56 (2011) 2895-2905.

[34] R. Fukui, Y. Katayama, T. Miura, J. Electrochem. Soc., 158(9) (2011) D567-D572.

[35] J. Mostany, J. Mozota, B.R. Scharifker, J. Electroanal. Chem., 177 (1984) 25-37.

[36] G.A. Gunawardena, G.J. Hills, I. Montenegro, Electrochim. Acta, 23 (1978) 693-697.

[37] M.-H. Yang, I.-W. Sun, J. Appl. Electrochem., 33 (2003) 1077-1084.

[38] D. Grujicic, B. Pesic, Electrochim. Acta, 50 (2005) 4426-4443.

[39] J.F. Moulder, W.F. Stickle, P.E. Sobol, K.D. Bomben, Handbook of X-ray Photoelectron Spectroscopy, Perkin-Elmer Corp., Eden Prairie, MN (1992).

9

Mechanical Property Evaluation of Electrodeposited Nanocrystalline Metals by Micro-Testing

Takashi Nagoshi, Tso-Fu Mark Chang and Masato Sone

Abstract

Electrodeposition is a very important technology in the fabrication of micro-components for micro-electro-mechanical systems (MEMS) or integrated circuits. Evaluations of the materials used in these devices as 3D components should be conducted using micro-sized specimens due to the sample size effect on the practical use of the components. Nanocrystalline metals could be deposited using an electrodeposition method with supercritical CO_2 emulsion. Our experiment on the micro-specimens provides information on micro-mechanical testing of electrodeposited metals including the effect of sample size, grain size, and anisotropic structures on mechanical properties. In this chapter, recent studies on crystal growth in electrodeposition of metals and its evaluation using micron-sized testing will be presented.

Keywords: Electrodeposition, metal, texture, microstructure, strength, micro-testing

1. Introduction

1.1. Microstructure of electrodeposited metal

Micro-electro-mechanical systems (MEMS) devises are usually fabricated using film deposition process. Deposition of metallic film can be classified into two major categories, one is the dry process and the other is the wet process. Dry process involves the use of gas or metallic vapor for deposition or directly deposit metal atoms on the surface of a substrate by sputtering. The process is simple, but the deposition rate is very slow, such as sub-nm to several nm in a second [1, 2]. Thus, the deposition is limited to sub-μm-scale structures in industrial applications and not favorable for fabrication of MEMS components that require structural support.

As one type of the wet process, electroplating has been used in the industrial fabrication of MEMS and integrated circuit (IC) thanks to the fast deposition rate, low production cost, and simple operation [3, 4].

In recent years, the miniaturization of MEMS and IC merges into a nano-scale regime, also called nanotechnology. The component size or wire width of the structures reach several tens of nanometers. In the filling of small gaps, some problems arise as schematically shown in Figure 1. Since liquid solutions are used in the wet process, when the substrate is not covered with the electrolyte or hydrogen gas, bubbles evolve in the reaction adsorbed on the substrate, the area not having contact with the electrolyte would lead to formation of voids and pin-holes [5]. To accomplish successful gap-filling, we have developed an electroplating method with supercritical CO_2 emulsion (EP-SCE) [6, 7]. In this system, micelles formed with surfactant encapsulates supercritical CO_2. These micelles randomly bounce on the surface of the cathode and desorb evolved hydrogen bubbles. In reaction areas where the micelle is in contact, deposition must be stopped and eventually resume with bulk concentration of electrolyte. These features, which are called periodic plating characteristics (PPC) [8, 9], will contribute to the gap or hole filling in EP-SCE and attained Cu filling of a hole 70 nm in diameter and 350 nm in depth as shown in Figure 2 [10, 11]. Results of the metal deposition using EP-SCE are summarized in Table 1. On each metal deposition, electrolytes with the same base were used while the EP-SCE contains additional surfactant and emulsified with supercritical CO_2.

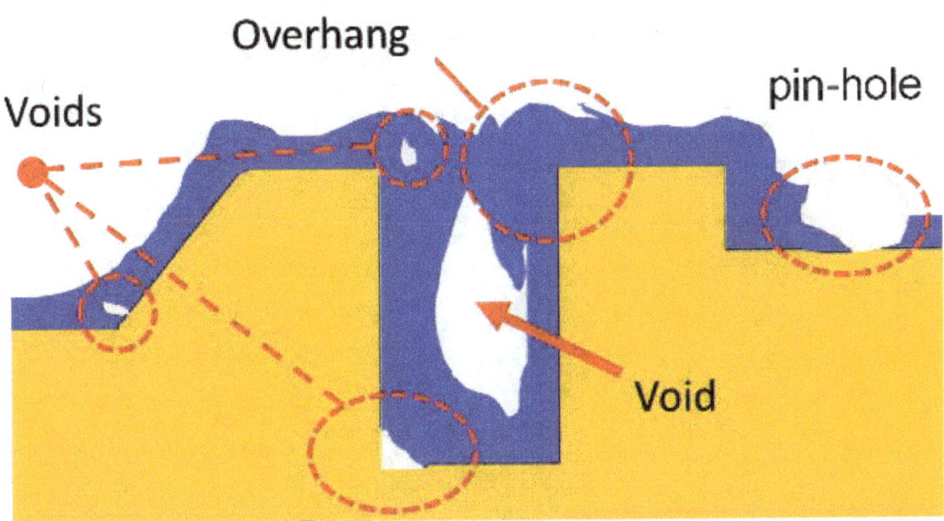

Figure 1. Failures found in filling of gaps with electrodeposition.

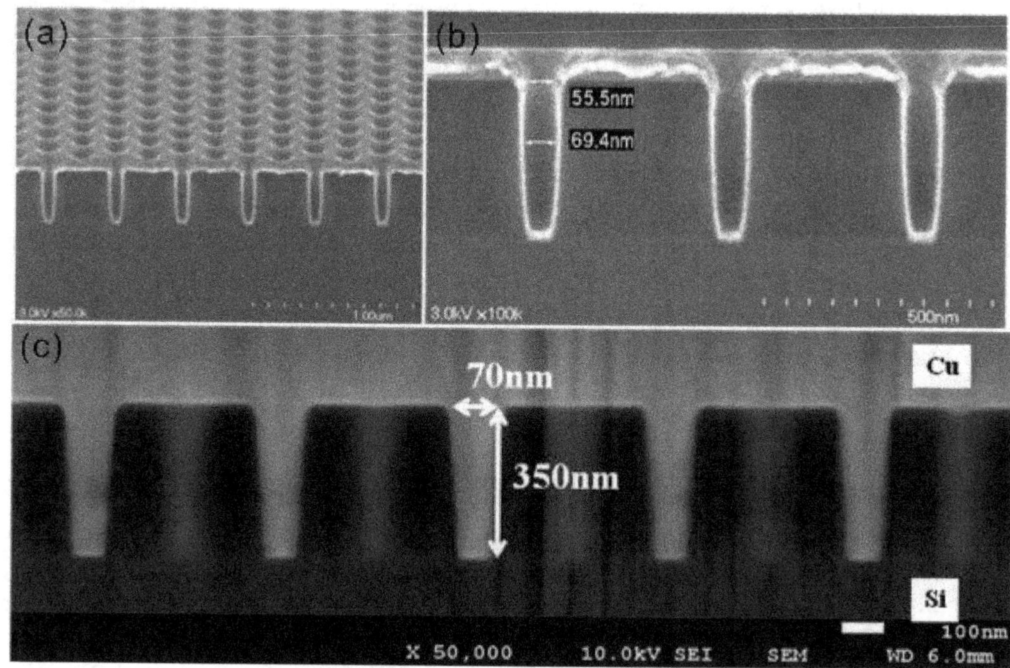

Figure 2. Cross-sectional SEM images of (a) Hole test element group (TEG) with holes of 70 nm in diameter and aspect ratio of 5, (b) expanded image of (a), and (c) TEG filled with Cu electroplated in EP-SCE with copper particles.

Plating method	Properties	Nickel [12, 13]	Copper [14]	Tin[15]
EP-SCE	Structures	Equi-axial grains	Equi-axial grains	Equi-axial grains
	Grain size	8 nm	100 nm	50 μm
	Impurities	2 at% of carbon	0.1 at% of carbon near substrate	Not detected by XPS
	Preferred orientation	No	-	-
CONV	Structures	Columnar grains	Polycrystalline with high density of twin boundary	Equi-axial grains
	Grain size	4.4x0.8 μm	1 μm	60 μm
	Impurities	Trace	0.1 at% of carbon near substrate	Oxygen as oxides
	Preferred orientation	<110> fiber texture	-	-

Table 1. Electrodeposited metals with EP-SCE and conventional (CONV) method

1.2. Mechanical property of electrodeposited metal

Micro-sized components used in MEMS fabricated by electrodeposition need to be tested to ensure device lifetime or tolerance to mechanical damage. MEMS are made up of components below 100 µm in size. For example, MEMS-based accelerometers or gyroscopes are widely used in cell phones, gaming consoles, and location-based devices. Some examples shown in Figure 3 are optical switch (3a) and gyro sensor (3b). Micro-components used in MEMS such as micro-spring, bending beams, and structural support of MEMS suffer from mechanical straining and need suitable mechanical properties. For mechanical property evaluation on such materials, conventional indentation or wear test is insufficient. Moreover, when the sample size comes to micro-scales, the classical physics are not always useful. Sample size effect, which will be described in a later section, emerges. Thus, a micro-testing method with a specimen whose sample size is in the same scales as the actual MEMS components is needed.

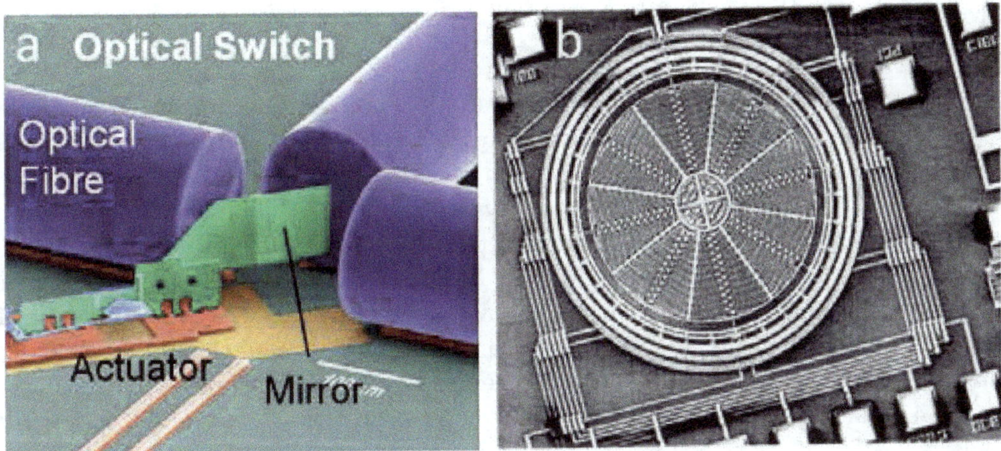

Figure 3. Examples of MEMS devises (a) optical switch and (b) gyro sensor.

1.3. Mechanical properties of small scale materials

The strength of metals has been shown to increase with decreasing sample size, known as "smaller is stronger" and also referred to as size effect [16]. For example, thin film experiments, including wafer curvature [17] bulge tests [18] and MEMS-based tests [19], have consistently shown an increase of strength with decreasing sample size. The majority of research investigating the size effect of metals by means of the micro-testing technique has focused on single crystalline metals [20, 21, 22, 23, 24, 25] and metallic grasses [26, 27, 28, 29]. Only limited data exists for metals with microstructures. Rinaldi et al. [30] and Jang and Greer [31] have investigated nickel nanocrystalline nano-pillars. However, their trends for strength as a function of sample size show opposite trend: Jang and Greer observe a "smaller is weaker" power-law dependence in nanocrystalline Ni; while Rinaldi et al. report very scattered results with slight strengthening with decreasing sample size for 30 nm-grained nanocrystalline Ni.

Thus, currently the experimental findings for sample size effect on nanocrystalline materials are inconclusive, and there is much uncertainty with respect to the deformation mechanism and combined effect of different size effects. Sample size effect on polycrystalline pillar has great interest in industries owing to miniaturization of MEMS devices reaching to its component scales at sub-micro or nano regime.

2. Electrodeposited Nickel

2.1. Ni film in conventional electrodeposition

2.1.1. Microstructural development of Ni film

Crystal growth in conventional Ni electrodeposition is investigated using additive-free Watt's bath. The substrate was films of pure Cu annealed at 673 K for 1 hour in vacuum. Ni layers with a thickness of about 25 μm were electrodeposited under agitation with magnetic stirring bar.

Microstructure of film cross-sections were evaluated by a scanning electron microscope (SEM) equipped with an electron backscatter diffraction pattern (EBSD) detector. The orientation map shown in Figure 4a was overlaid with a grain boundary map where $\Sigma 3$ boundaries were colored yellow. The fraction of grain boundary was summarized in Figure 4b as low-angle grain boundaries (LAGB) with misorientation between 2° and 15°, high-angle grain boundaries (HAGB) with misorientation above 15°, and $\Sigma 3$ boundaries. Fine columnar grains with diameter of around 100–200 nm have grown toward the film surface as shown in Figure 4a. Most of the grain boundaries dividing columnar grains are $\Sigma 3$ boundaries, corresponding to twin boundaries, although the stacking fault energy of Ni is relatively high [32]. High density of twins, 42% among HAGB in this film, was also reported in various kinds of electrodeposited metals [33, 34, 35].

Transmission electron microscopy (TEM) observations in the vicinity of the interface between the electrodeposited Ni and Cu are shown in Figure 5 as bright field TEM image in 5a and schematic illustration corresponding to the TEM image in 5b [12]. In the epitaxial region, only about 100-nm thick contains high density of dislocations. Misfit strain of 2.5% due to the difference in the lattice constants of Cu (3.615 nm) and Ni (3.524 nm) arises between the electrodeposited Ni layer and Cu substrate. We consider that the observed dislocations were introduced to accommodate this misfit strain. Since the orientation of the epitaxial region is almost the same as that of the Cu substrate, an epitaxial region was formed to minimize the interfacial energy between the electrodeposited Ni layer and Cu substrate. The misfit strain, however, restricts the growth of the epitaxial region up to about 100 nm. Ahead of the epitaxial regions, columnar grains without an orientation relationship with respect to the Cu substrate form are shown in Figure 4. TEM image of the Ni film 1.4 μm away from interface is shown in Figure 6. The thickness of the twin is about 10 nm, which is substantially narrower than the twins observed in EBSD shown in Figure 6. From the TEM observation, these narrow twins with a thickness of about 10 nm were frequently observed while they cannot be detected in

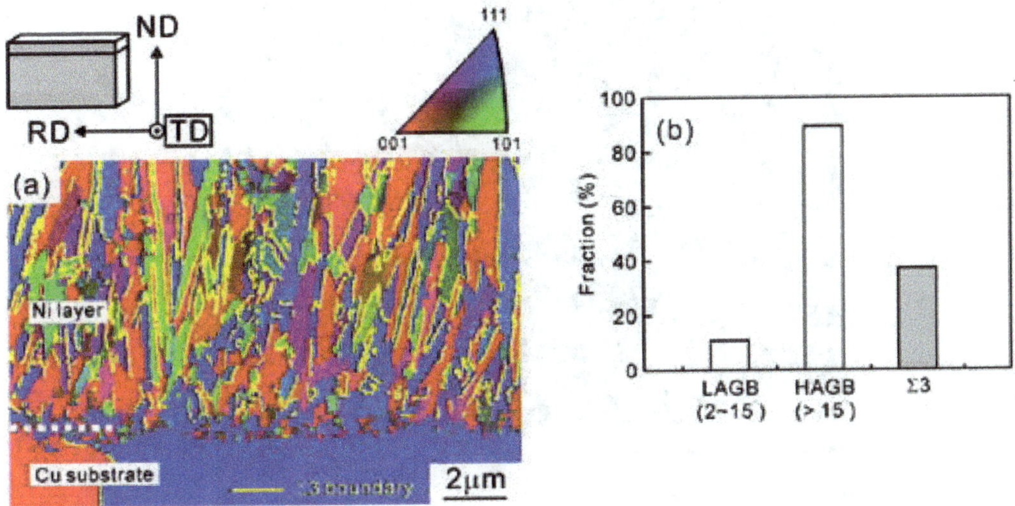

Figure 4. (a) Cross-sectional orientation maps of electrodeposited Ni layer. Σ3 boundaries are indicated as yellow lines in the figure. (b) Fractions of boundaries shown in (a) with different nature. Boundaries with misorientation between 2° and 15° classified as low-angle boundaries (LAGB) and misorientation angle above 15° are high-angle boundaries (HAGB). Σ3 boundaries that are HAGB satisfy the Brandon criterion (Σ3).

EBSD. We suppose that the twins in the electrodeposited Ni layer formed during electrodeposition reaction in the course of lateral movement of {111} facet shown in Figure 5a. Change in stacking sequence occurred at {111} facet could turn into a nucleus of twin grain. Fujiwara et al. [36, 37, 38] also proposed a similar model of twin formation parallel to the growth direction during the melt growth of Si.

2.1.2. Evaluation of anisotropic columnar grains in electrodeposited Ni by micron-sized cantilever

Mechanical properties of Ni film electrodeposited in conventional Watt's bath was evaluated using micro-cantilever. Cantilevers were milled out of the Ni film using focused ion beam (FIB) with different beam directions to investigate anisotropic mechanical properties. Micro-bending test had been used to investigate size dependent effect [39], local mechanical properties of lath martensite [40], and anisotropic fracture toughness of the NiAl single crystal [41]. In the cantilever specimen, deformation takes place at the fixed end on the tension at the upper side and compression at the lower side. Therefore, site specific influence of structure on the mechanical properties and deformation behavior can be examined.

Two cantilevers of 10 x 10 x 50 μm were fabricated with beams parallel and perpendicular to growth direction. As shown in Figure 7, columnar grains are aligned to the growth direction indicated in black arrow. The bending test had been carried out by indenting at the cross mark on the beam that is 40 μm away from fixed end. Force displacement curve in Figure 8 shows clear increase in the bending stress of the parallel cantilever. Maximum stress obtained was 2, 080 and 1, 582 MPa for the parallel and perpendicular cantilevers, respectively, with respect to the growth direction [42].

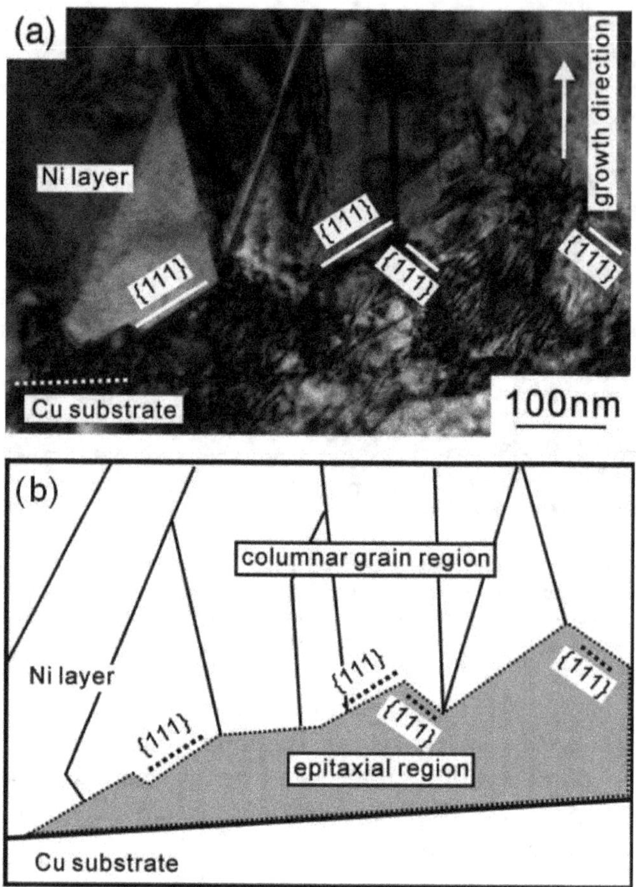

Figure 5. (a) Bright field TEM images showing the cross-sectional microstructure near the interface between the electrodeposited Ni layer and Cu substrate and (b) schematic illustration of (a).

In the deformation of fcc metals in which the slip system is {111} <110>, dislocation easily glides along the (111) twin plane. Ni film electrodeposited in conventional Watt's bath had high density of (111) coherent twin boundary parallel to growth direction. For the cantilever with parallel columnar grains, dislocations can move away from the stress concentration area. On the other hand, for the cantilever with perpendicular columnar grains, dislocations stop at the neutral plane. Longer slide distance of dislocations means effective stress relaxation at the fixed end, which enhances the apparent strength of the cantilever.

2.2. Ni film electrodeposited in supercritical CO_2 emulsified electrolyte

2.2.1. Nano-structured grains in Ni electrodeposited in supercritical CO_2 emulsified electrolyte

EP-SCE was conducted to deposit Ni on Cu substrate. Agitation with addition of surfactant, polyoxyethylene lauryl ether ($C_{12}H_{25}(OCH_2CH_2)15OH$) enables an electrolyte and supercritical

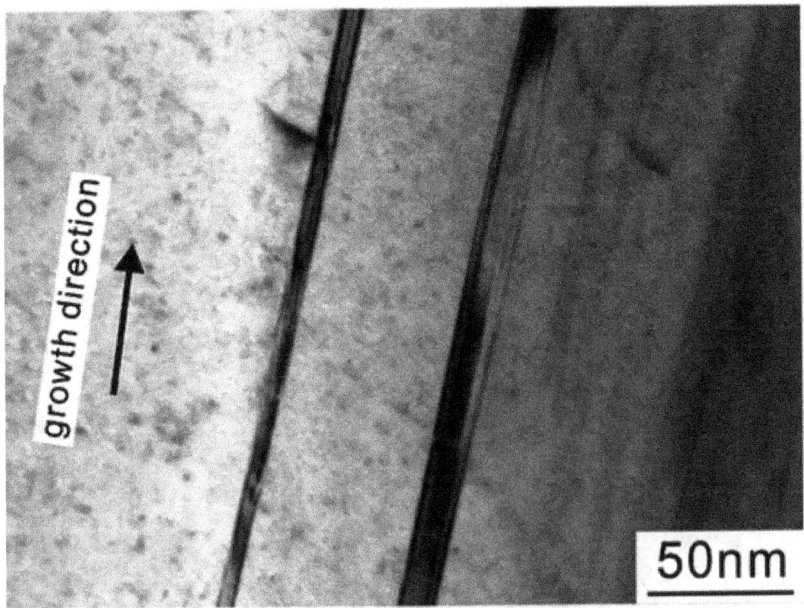

Figure 6. Bright field TEM image showing the microstructure within the columnar grain at 1.4 μm away from the interface between the electrodeposited Ni layer and Cu substrate.

CO_2 to form emulsions with CO_2-in-water (C/W) type micelles [43]. Additive-free Watt's bath was emulsified with CO_2. Electrodepositions on Cu substrate were conducted at different pressure of 6, 10, 15, 20 MPa including conventional electroplating (CONV).

The TEM bright field image shown inFigure 9is nickel film fabricated by EP-SCE at 15MPa viewed from growth direction [13]. Figure 10 shows average grain size of the films fabricated at different pressures based on measurement of around 100 grains in TEM. Equi-axial grains were found in all samples viewed from the sample surface and cross section. Grain refinement can be achieved by pulsed plating characteristics as stated by Chang et al., where the bouncing micelles promote nucleation by inhibiting grain growth [9]. Density of the CO_2 inside micelles increases with the increase in pressure, especially near the transition point from gas to supercritical phase. Change in density of the dispersed phase in emulsion will have an effect on the micelle structure and dispersion conditions. Thus, the change in grain size by different applied pressure was expected.

The impurity content of the film obtained from glow discharge optical emission spectroscopy (GDOES) should be noted. The carbon content could contribute to the grain refinement, which is shown in Table 2 including impurities of boron and oxygen. Comparing with CONV-plated nickel with and without surfactant, some impurities derived from surfactants were detected. However, in the emulsified state, surfactants consumed by the formation of micelles and the number of surfactants adsorbed and involved in the film could decrease. Similar concentration levels of oxygen among CONV- and EP-SCE-plated films imply surfactant extinction at the

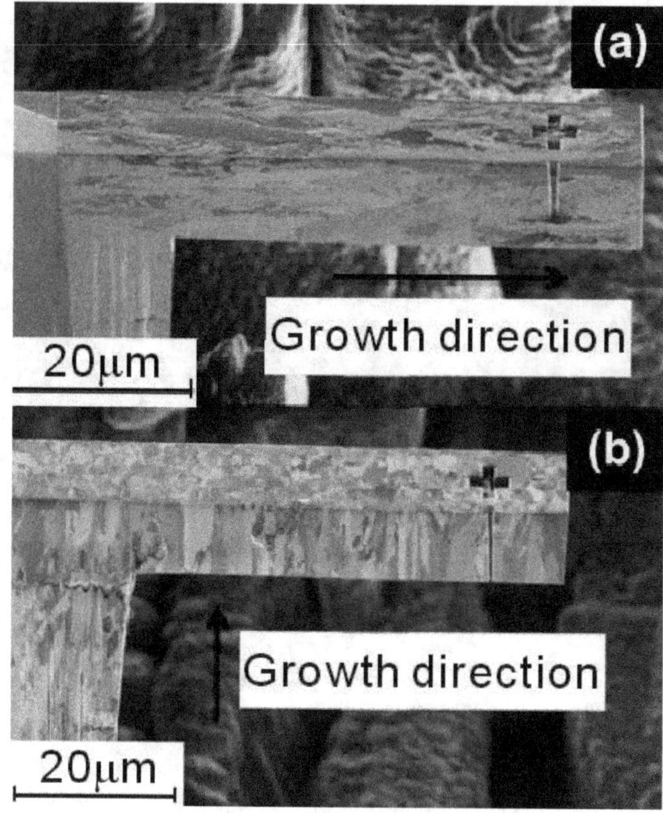

Figure 7. FIB images of the cantilevers fabricated by FIB. Beam directions are (a) 0° and (b) 90° with respect to growth direction indicated in black arrow.

surface when surfactants are mainly composed of carbon and oxygen. These results show that carbon derives from CO_2 dissolved in an electrolyte.

Deposited film	Concentration (at %)		
	Boron	Carbon	Oxygen
CONV without surfactant	0.000785	0.015	0.027
CONV with surfactant	0.00564	0.46	2.54
EP-SCE at 6MPa	0.000910	1.91	0.0220
EP-SCE at 10MPa	0.00114	2.43	0.0355
EP-SCE at 15MPa	0.00114	2.61	0.0305
EP-SCE at 20MPa	0.00102	2.17	0.0153

Table 2. Concentration of impurities in deposited film

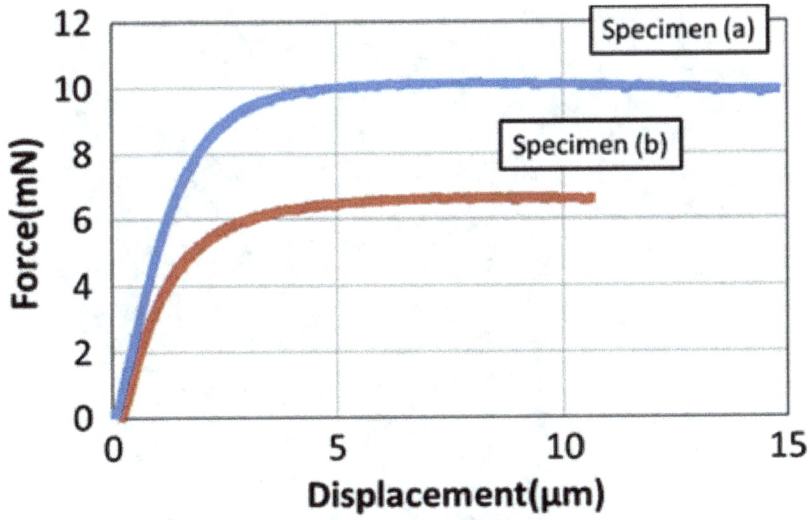

Figure 8. Load-displacement curve of the bending test. Specimen (a) and specimen (b) have beam directions of 0° and 90° with respect to growth direction, respectively.

Codeposition of carbon in electrodeposition has been reported by Chung and Tsai in supercritical CO_2 deposition [44] and Yamachika et al. using Au-Ni bath with citric acid [45], indicating that carbon originates from CO_2 and citric acid, respectively. Although carbon dioxide is chemically stable, production of hydrocarbons from carbon dioxide had been demonstrated in electrochemical reduction [46]. When the reduction of CO_2 dissolved in electrolyte occurred, carbon involved in the deposited nickel film results in Ni-C alloy deposition. In alloy electrodeposition where the foreign elements are incorporated as interstitials, crystalline size decreased when composition of foreign elements increased [47]. The alloying elements inhibit grain growth in the deposition process leading to the formation of amorphous film [48, 49]. The EP-SCE nickel also showed decreased grain size with increasing carbon content. Lattice constant observed in the X-ray diffraction (XRD) showed expansion corresponding to 2% of the solute carbon. Supersaturation of carbon solute in nickel deposits, where carbon can easily segregated on the grain boundary [50] and inhibit grain growth that results in nanocrystalline Ni deposits.

2.2.2. Mechanical property of nanocrystalline Ni evaluated by micro-compression test

Mechanical properties of nanocrystalline EP-SCE nickel, including the effect of grain size and sample size, were investigated using a micro-compression test. Non-tapered micro-pillars were fabricated by FIB from the region analyzed by the EBSD technique. Sequences of pillar fabrication are illustrated in Figure 11 with corresponding scanning ion microscope (SIM) images. In the course of fabrication, we first made a pillar that has the thickness of thin plate of around 100 μm and was shaped as shown in Figure 11a. Using irradiation of 45° from the thin plate, we made a small pillar from the bigger one as shown in Figure 11b. The ion beam from the side of the specimen allowed the fabrication of a pillar with uniform dimensions (non-

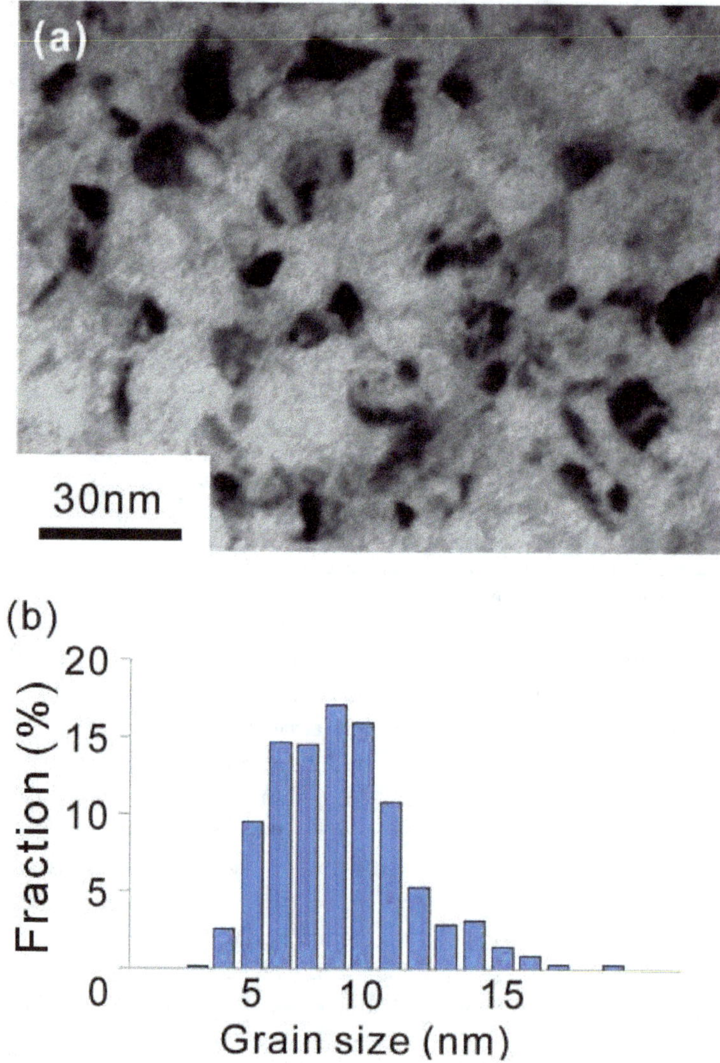

Figure 9. (a) Plan-view TEM bright field image of EP-SCE nickel and (b) grain size distribution from more than 500 of grains observed in TEM.

tapered, non-filleted). Finally, we milled each side of the pillar at a tilt angle of ±2.3° with 400 pA ion beam to minimize ion bombardment damage. A 20-μm square cross-section pillar fabricated from nickel film plated in EP-SCE at different pressures was employed to investigate grain size effect on plated nickel.

Micro-compression tests of nanocrystalline Ni with different grain size were conducted and the deformed pillars were observed as shown in Figure 12. All the pillars were deformed by broad shear banding crossing through top to bottom. However, the nickel film electrodeposited at 6 and 20 MPa, which has a larger grain size of 15 nm, had shown notable difference of

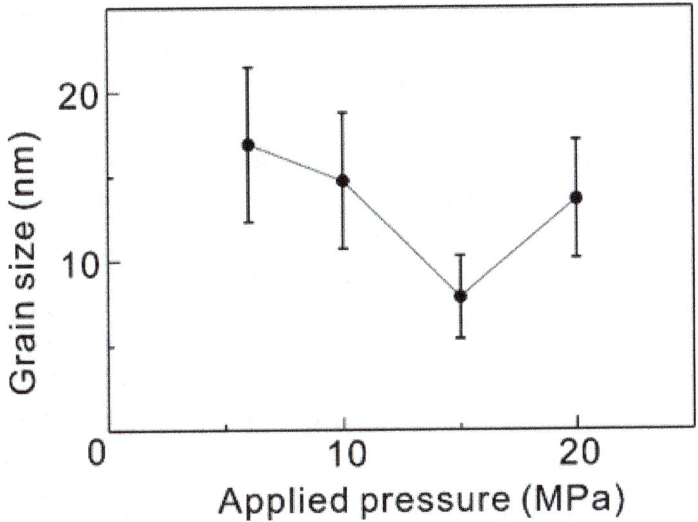

Figure 10. Grain size of the films electrodeposited at different pressures evaluated by TEM.

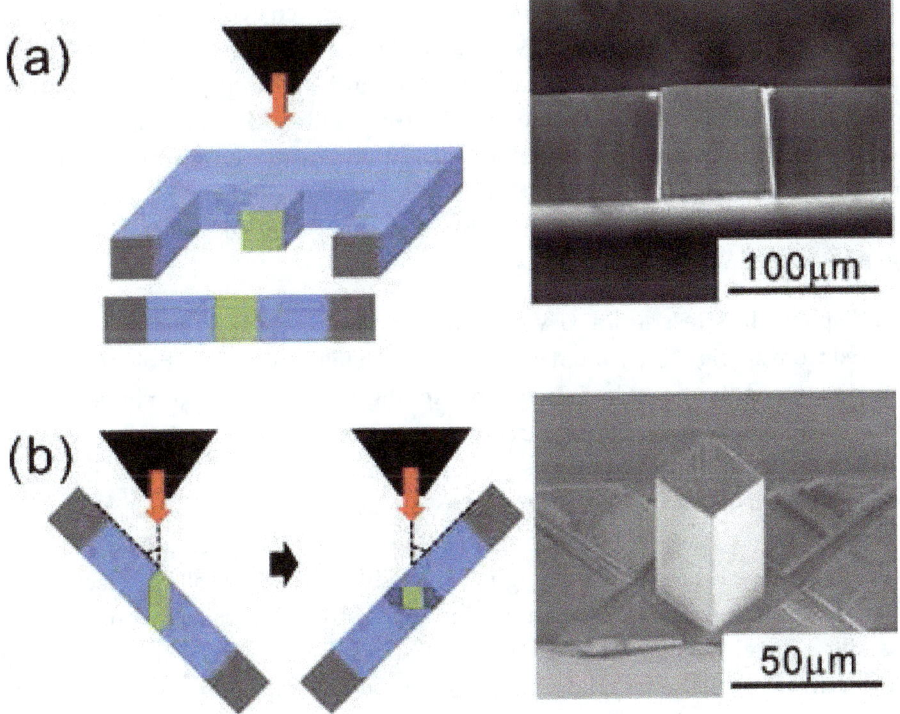

Figure 11. Procedures of making a compression pillar and SIM images showing the pillar after the procedure: (a) perpendicular irradiation and (b) irradiation from 45° from the thin plate.

bulging on the pillar surface as shown in Figure 12a and 12c. Micro-compression test results are shown in Figure 13 as true stress–true plastic strain curves. No clear relationships were found between deformation morphology and stress-strain behavior. Highest maximum stress of 3, 500 MPa was observed in the film electrodeposited at 15 MPa, which had the smallest grain size of 8 nm. The contribution on strengths of supersaturated carbon interstitials was calculated based on Fleischer formulation [51]. A 7.9 MPa of strengthening responsible for 2 at% of interstitial carbon is negligible compared to 3.5 GPa of maximum stress. The high strength in EP-SCE nickel could be due to a suppressed Hall-Petch breakdown. The strength of nanocrystalline Ni with grain size of 8 nm is on the extrapolated Hall-Petch slope from nickel alloy as shown in Figure 14. One-third of hardness values of pure nickel [52, 53, 54] and Ni-W alloy [55] are included in the Hall-Petch plot assuming Tabor relation [56] as well as 0.2% yield stress in the present micro-compression test. The suppression of Hall-Petch breakdown is consistent with a literature reported by Schuh et al., the hardness of Ni-W alloy has fallen at 8–9 nm and suggested that the alloying with tungsten has suppressed the Hall-Petch breakdown [55]. They concluded that the slow diffusion of tungsten in nickel increase required stress for activation of Coble creep and grain boundary sliding. Present electrodeposited nickel have 2.6 at% of carbon impurities and 2.0 at% of interstitial carbon, and the rest of the carbon could be segregated at the grain boundaries. Contrary to the tungsten in nickel, carbon has very high diffusivities in nickel via interstitial site diffusion. Yin et al. reported an effect of interstitials on creep deformation via Coble creep and grain boundary sliding [57]. On that literature, interstitially dissolved atoms are reported to effectively enhance creep resistance. Hall-Petch breakdown represents a transition of deformation mechanism from dislocation-mediated to grain-boundary-mediated. Thus, if the grain boundaries are reinforced by impurities, dislocation motion dominates the deformation at much smaller grain size and results in suppressed Hall-Petch breakdown [58].

Figure 15 displays the stress strain curves of pillars with sample size ranging from 5 μm to 30 μm prepared from single crystal and nanocrystalline Ni [59]. Due to multiple slip glides across the pillar or toward the base, large work hardening was observed in single crystal pillar compression. This result is similar to the compression of <111> oriented nickel pillar by Frick et al [20]. They observed dislocation lines throughout the pillar and base of the pillar, which indicates dislocation interaction by multiple slip and accumulation of dislocation at the pillar base. Thus, the stress needs to activate dislocation source inside the pillar increased with increasing strain. The following softening is believed to be a result of macroscopic shear by the activation of different slip systems due to increased stress. The deformation mechanism of nanocrystalline metal is believed to be a grain boundary process, such as grain boundary sliding or grain rotation. Considering the deformation mechanism, activation of dislocation source, which believed to be a possible explanation of size-dependent strength [16], did not play a main role in plastic deformation of nanocrystalline materials. Thus, the sample size effect on electrodeposited nanocrystalline Ni was not expected. However, the micro-compression test shows obvious increase in both yield stress and flow stress with decreasing sample size from 20 μm to 5 μm.

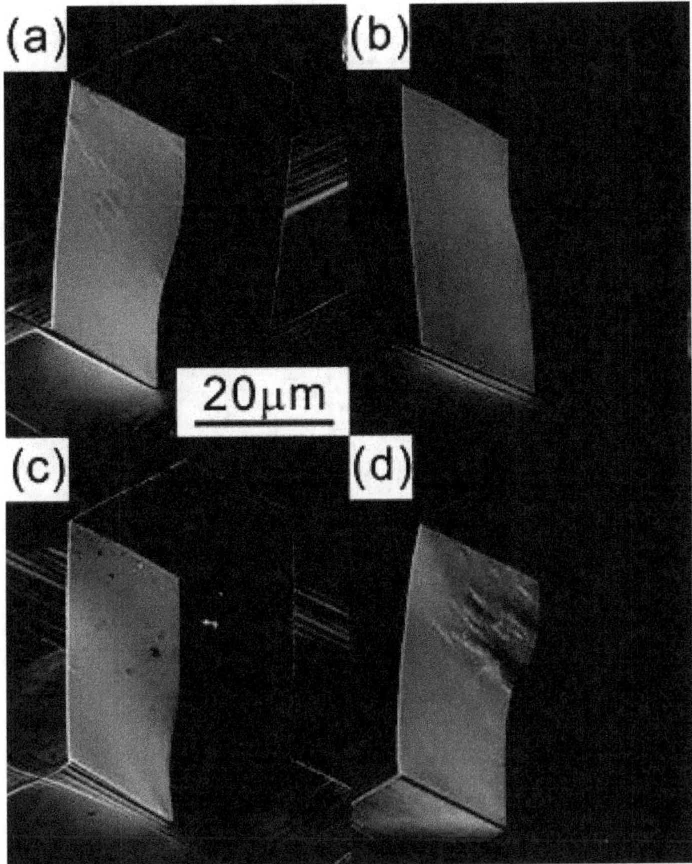

Figure 12. SEM images of the deformed pillars (a) 6 MPa, (b) 10 MPa, (c) 15 MPa, and (d) 20 MPa.

The stress as a function of pillar diameter for single crystal and nanocrystalline Ni was shown in Figure 16 [59]. The scaling exponent of 0.25 for the peak stress of single crystal Ni is small compared to 0.64 observed by Dimiduk et al. using non-tapered single crystal Ni (269) [21]. This can be explained by the change in dislocation mechanisms inside the pillars since different loading directions were taken. Relatively large sample sizes and multiple slips in the present work provide dislocation pile-ups and interaction of dislocation results in more dislocations inside the pillar that hinders sample size effect; note that the strength taken as a peak stress in stress strain curve is due to the uncertainty of yield point. For nanocrystalline Ni, which believed to deform without dislocation activation, the scaling exponent of 0.057 was observed. Although the exponent is quite low, strength obviously increased from 3.6 GPa to 4.1 GPa in 30 to 5 μm pillars. Reports on size effect of nanocrystalline materials are very limited and still controversial. Rinaldi et al. found increased strengths with the decreasing diameter of pillars from 270 to 160 nm in diameter-fabricated from nanocrystalline nickel with 30 nm of grain size [30]. On the contrary, Jang and Greer demonstrated size-induced weakening in 60 nm grained nickel nanopillars with a diameter between 3, 000 nm and 100 nm [31]. This contradiction can

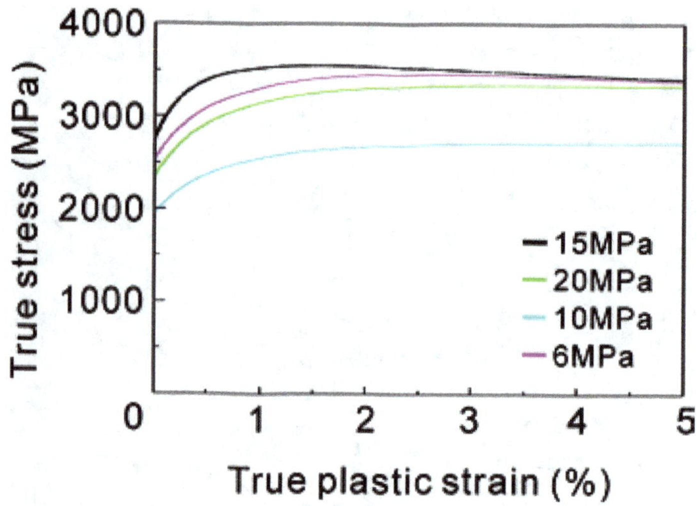

Figure 13. True stress–true plastic strain curves of a micro-pillar from films electrodeposited at different pressures.

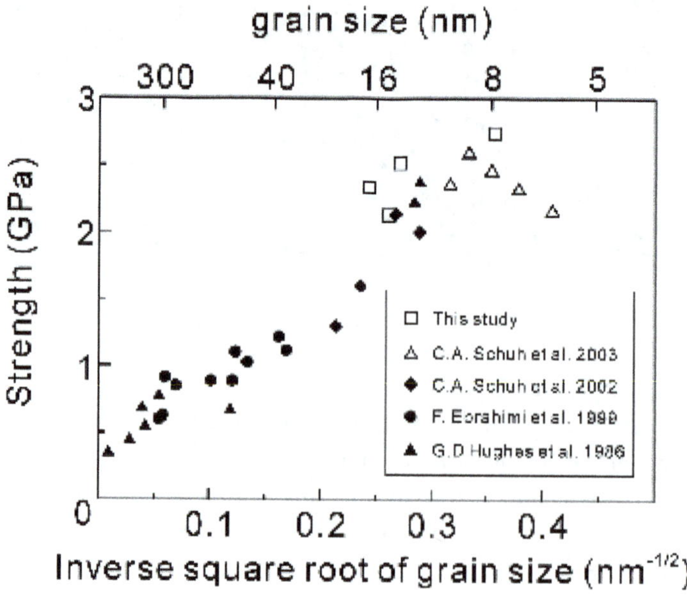

Figure 14. Hall-Petch plot representing 0.2% yield stress as well as one-third of hardness value found in literature.

be explained based on the deformation mechanisms. Jang observed a transition of deformation mechanisms from dislocation-mediated to grain-boundary-mediated, while grain-boundary-mediated deformation supposedly dominated in the present work. Grain boundary sliding is reported to involve several grains in formation of micro-shear band along the grain boundaries. Sums of these shear band formation will cause macroscopic yield in the present micro-compression test. Microscopic strains can generate on the large area of grain boundaries, which

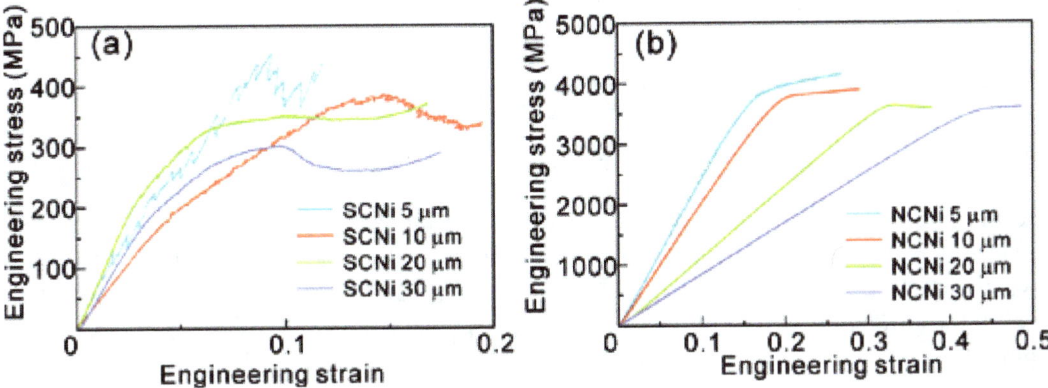

Figure 15. Engineering stress-strain curves for micro-compression of (a) single crystal Ni pillar and (b) nanocrystalline Ni pillars with different pillar sizes.

lies on the same plane with larger shear strain. This corrective motion of grain boundaries has been known as cooperative grain boundary sliding (CGBS) [60]. Zerin and Mukherjee observed bimodal distribution of sliding offset length on each CGBS event, which indicates breaking up of large sliding grain block into small grain blocks by secondary CGBS operation. This is observed with increase of strain, i.e., work hardening. CGBS events could initiate from the flat segment of grain boundaries and the number of these segments decreased when the sample size becomes smaller. Larger samples have segments of grain boundaries with longer distance to sliding direction and can deform with smaller stress. This is in good agreement with the change in exponent with increased strain, where large grain blocks in large samples can break up while small samples deform by CGBS with small grain blocks.

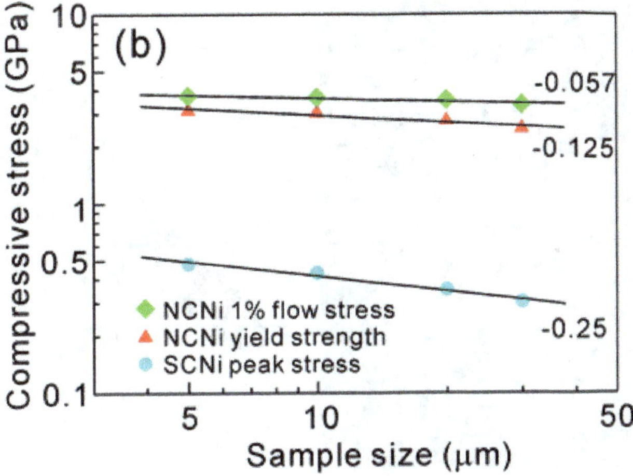

Figure 16. Strengths in micro-compression, peak stress for single crystal Ni, 1% flow stress, and 0.2% offset stress for nanocrystalline Ni are shown in a double-logarithmic graph. Solid lines represent power law fittings and each exponent is shown on the right.

3. Electrodeposited copper

3.1. Cu film electrodeposited in conventional electrolyte

3.1.1. Microstructure and self-annealing of Cu

Cu films were electrodeposited on Cu substrate in sulfate bath with and without additives. Current density of 2A/dm^2 and 10 A/dm^2 were used to deposit around 20 μm of Cu film. Investigation on microstructure of the electrodeposited Cu film was conducted immediately after the electroplating process and continued until self-annealing was observed. Cross-section EBSD was employed to measure grain size.

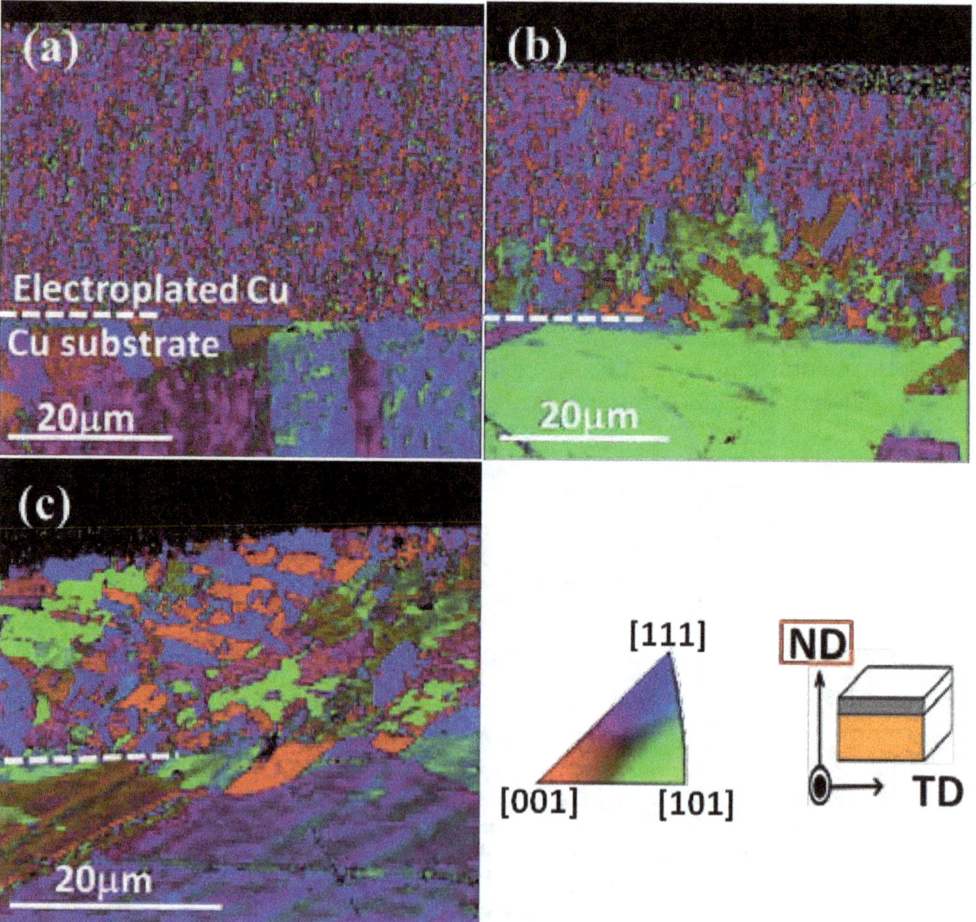

Figure 17. Cross-sectional EBSD maps for sample normal direction of self-annealed plated Cu films on Cu substrate at current density 100 mA/cm^2 (a) after 24 hours of incubation, (b) after 140 days of incubation, and (c) after 250 days of incubation. Dashed line indicates the interface.

Cross-sectional orientation maps obtained from EBSD analysis on Cu film deposited without additives are shown in Figure 17 [35]. Initial grain size was measured as 330 nm and enlarged to 860 nm after 250 days of incubation in vacuum. On the other hand, Cu film deposited with additives had large grain size near 1µm, which is obviously different from one of the Cu film that was as-deposited without additives and more similar to the film after self-annealing. The Cu film electrodeposited with additives may had been self-annealed soon after electrodeposition before characterization of structures. Many factors influence self-annealing, such as film thickness, incorporated impurities, and initial crystal structures [61, 62, 63]. Additives are usually used to smoothen the film and are known to refine initial grains while inhibiting grain growth during electroplating. Although the incorporated impurity inhibits recrystallization and slow down self-annealing, the significantly decreased grain size accelerates it and may have results in self-annealing within sample preparation for structure observation. Stangl et al. reported the effect of additives where the addition of a slight amount of additives cause nearly 8 times faster recrystallization at room temperature [61].

Table 3 shows the change in the fraction of grain boundaries with incubation time. All self-annealed microstructures in the electroplated Cu film contained a relatively high fraction of HAGB. Fraction of HAGB in the electroplated Cu film gradually decreased with the increase in incubation time. The increase in grain size results in decreased grain boundary area of HAGB, with lower boundary free energy in the materials. Therefore, the driving force of self-annealing of the electroplated Cu film at room temperature is boundary free energy of the high fraction of HAGB. This suggestion about driving force is supported by many researchers [62, 64]. In addition, the fraction of twin boundaries is increased by incubation time as shown in Table 3. Figure 18 shows the recrystallized area of Cu film after 140 days of incubation. Orientation of the individual grains are shown in Figure 18b as {110} pole figure with (111) plane, the sharing in each grain are marked with the dashed line. Grain A–E sharing (111) plane with each other, such as A/B, B/D, D/E, and B/C. Moreover, geometries of the twin plane and (111) plane seem identical, i.e., angles of grain boundaries between A and B, and D and E are close to the angles of (111) plane in the pole figure, which indicates these twin planes are coherent twin boundaries. The twin boundary, especially the coherent twin boundary has extremely low boundary free energy compared to HAGBs [64]. Practically, the fraction of HAGBs was found to decrease as self-annealing proceeds. Therefore, we suggest that the self-annealing was affected by the high fraction of HAGBs in the initial structure of the electroplated Cu film.

Incubation time	Fraction (%)		
	LAGB	HAGB	$\Sigma3$ (twin)
24 hours	9.60	90.40	25.52
140 days	17.18	82.82	29.68
250 days	37.46	62.54	29.22

Table 3. Fractions of grain boundaries

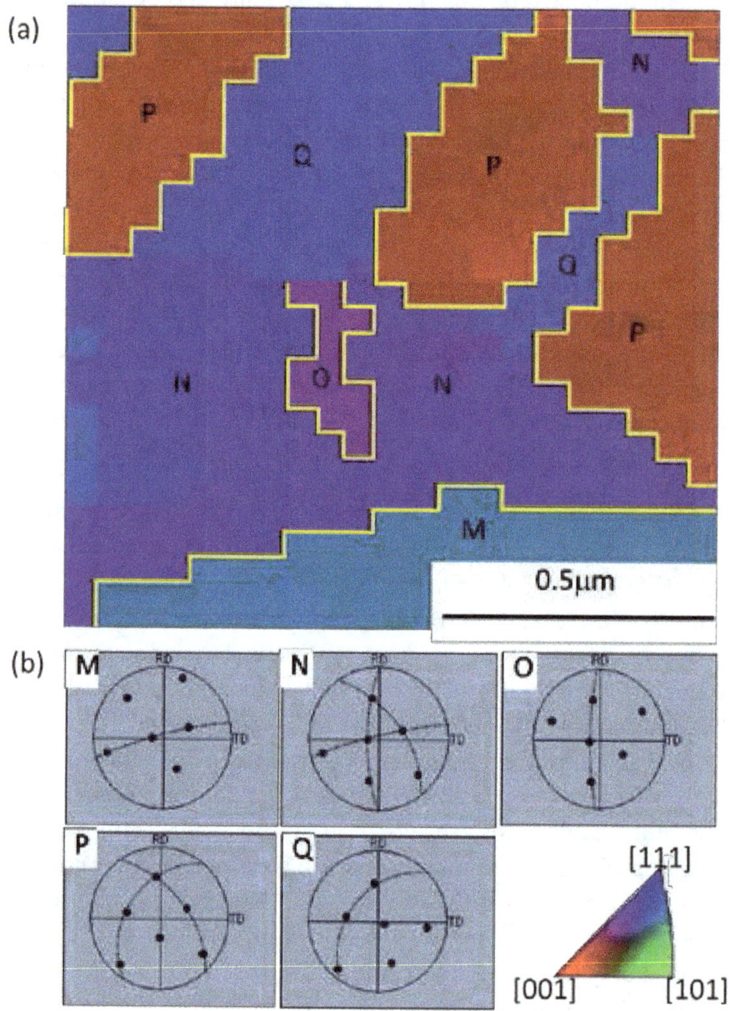

Figure 18. (a) Enlarged view of the cross-sectional EBSD map of plated Cu film after 140 days of incubation, and A to E in the map are equal to each grains. The yellow line shows twin boundaries. (b) {110} pole figures corresponding to the grain of A to E and the dashed line indicates the (111) plane.

3.1.2. Mechanical property of Cu electrodeposited in conventional electrolyte

The non-tapered micro-pillar was fabricated to evaluate mechanical properties and size effects of electrodeposited Cu films. The reduction in the number of grains within the sample may alter the mechanical properties of materials, which is one of the size effects [65]. In order to discuss the size effect, Cu films were electrodeposited using various current densities to change grain size whereas the size of all pillars is constant.

Table 4 shows average grain size, where measurement and calculation were made from more than 500 of grains observed in EBSD [66]. Grain boundary maps obtained from EBSD are shown in Figure 19. Grain size of the electrodeposited Cu film was smaller when higher current density was used. Square pillars with 20 μm on a side were fabricated using FIB from each deposited film. Compression test results are shown in Figure 20. The strength of the pillar increased with decreasing grain size corresponding to the well-known Hall-Petch relationship [67, 68]. Slight difference in flow stress in the compression test of two pillars plated at 0.5 A/dm² might be caused by the specimen size effect stated by Armstrong [69]. When the ratio between the specimen size (S) and average grain size (G), shown as:

$$\frac{S}{G} = \frac{Specimen\ Size}{Grain\ Size}$$

is small, deformations of the specimen depends on the individual grain in the sample. In extreme conditions, for a single crystal specimen where S/G ratio below one, yield stress is determined by the orientation of the single crystal. When the number of grains within the specimen becomes lower, it means smaller S/G ratio; the effect of orientation for each grain in deformation behavior become stronger. The S/G ratio of pillars of 0.5 A/dm² is smaller than pillars of 5.0 A/dm² or 2.0 A/dm² as shown in Table 4. Therefore, the deviation of flow curve in the compression test with the pillars with S/G ratio is smaller than 25 affected crystallographic orientations.

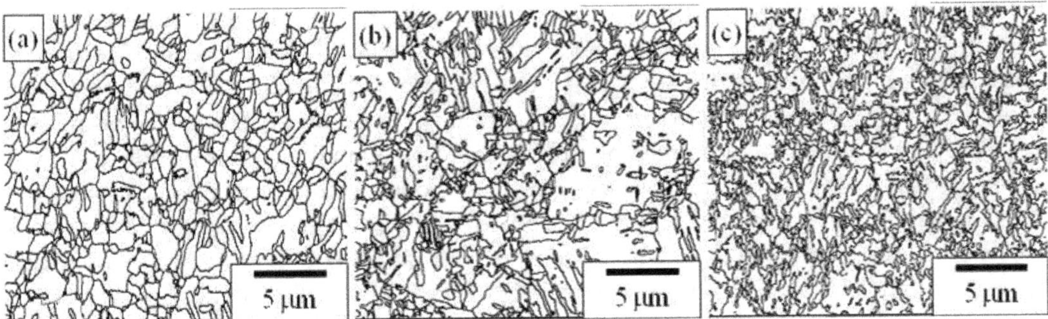

Figure 19. Grain boundary maps of electrodeposited Cu films. Cu film electrodeposited (a) at 0.5 A/dm², (b) at 2.0 A/dm², and (c) at 5.0 A/dm².

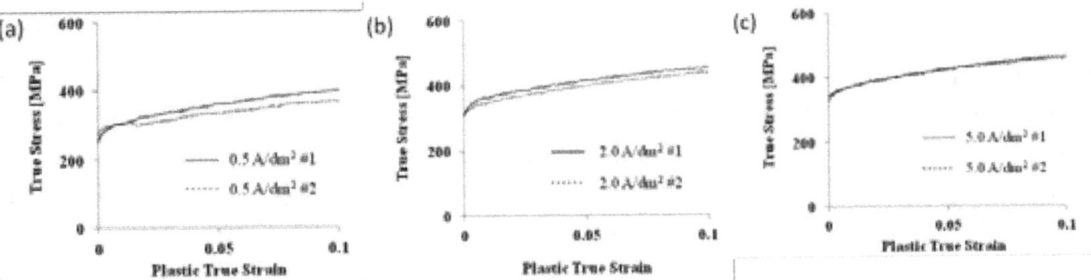

Figure 20. True stress–true plastic strain curves of the Cu pillars electrodeposited. Cu electrodeposited (a) at 0.5 A/dm², (b) at 2.0 A/dm², and (c): at 5.0 A/dm².

Current density [A/dm^2]	Grain Size [μm]	S/G ratio	Yield strength (0.2% offset) [MPa]
0.5	0.93	21.5	276
2.0	0.78	25.6	309
5.0	0.64	31.3	330

Table 4. Grain size, S/G ratio, and yield strength of the Cu pillars

3.2. Cu electrodeposited in supercritical CO_2 emulsified electrolyte

3.2.1. Nanostructure of electrodeposited Cu

Cu electrodeposition using EP-SCE with Watt's bath was conducted. Same high pressure apparatus in Ni deposition was used to deposit 10 μm of Cu film on Cu substrate. Figure 21 shows FIB images of deposited film [14]. Grain size of as-deposited Cu was significantly refined to 100 nm compared to CONV Cu. The cause of grain refinement could be corresponded to the one observed in Ni deposition. However, elemental analysis reveals very low level of impurities incorporated in the Cu film. Carbon impurity inside Cu film deposited in EP-SCE and CONV observed in GDOES is shown in Figure 22. Similar carbon distribution observed in CONV and EP-SCE Cu film indicates that the carbon impurity derives from an identical source. In both electrolytes, additives such as suppresser, accelerator, and leveler were used. These additives are adsorbed at the substrate and involved into the film during electroplating reaction. Thus, the CO_2 dissolved in the electrolyte used in EP-SCE had not reduced to carbon, which is different from Ni electrodeposition. More importantly, EP-SCE Cu showed significant microstructure evolution at room temperature as shown in Figure 21. Grain size of the film increased to 1 μm after two months of storage in a vacuum at room temperature. This is correspondent with self-annealing observed in Cu film deposited in conventional electroplating. Orientation maps shown in Figure 23 overlaid with HAGB and twin boundary in black and yellow lines. Structure change accompanied with the evolution of twin boundary was correspondingly observed with the one in the CONV Cu. High fraction of twin boundary, 70% of HAGB, should be the result of lowering boundary free energy in the process of self-annealing as discussed in Chapter 3.1.1.

3.2.2. Mechanical properties of nano-crystalline Cu electrodeposited in supercritical CO_2 emulsified electrolyte by micron-sized pillar

Mechanical properties of Cu before and after self-annealing was evaluated using micron-sized pillar fabricated using FIB. Figure 24 shows true stress-strain curve of Cu pillars fabricated from the films of as-deposited and two months after EP-SCE, including the results of CONV Cu [14]. A significant difference is observed in the mechanical strength among as-deposited EP-SCE and CONV. Engineering stress of the as-deposited EP-SCE pillar was about 300 MPa higher than the CONV pillar. And strength decreased by self-annealing to the same level of the CONV pillar. Deformation behavior is very similar in both self-annealed EP-SCE pillar and CONV pillar, in agreement to the identical microstructure observed. There is linear work hardening regime in the compression test of the CONV pillar and self-annealed EP-SCE pillar,

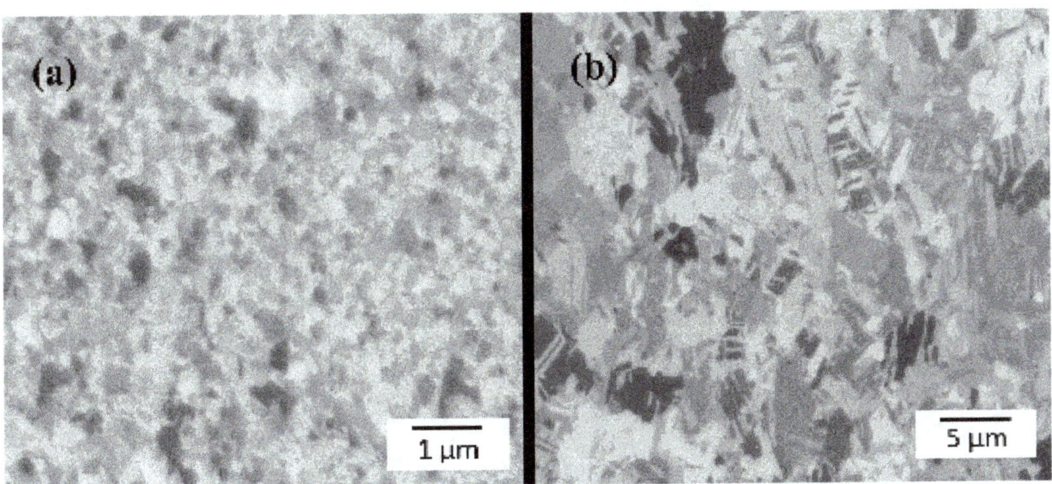

Figure 21. FIB images of Cu film obtained by EP-SCE (2.0 A/dm² for 120 min) (a) as electroplating (b) after two months.

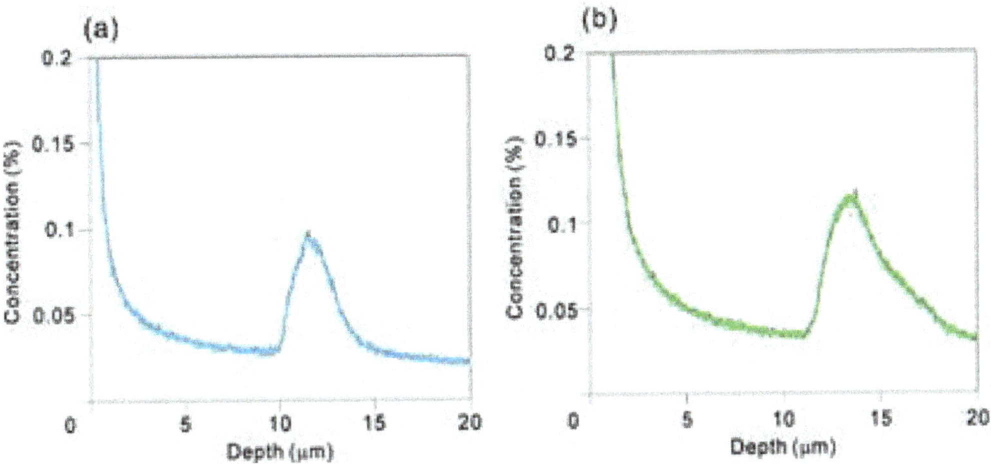

Figure 22. Carbon impurity concentrations in the copper film fabricated by (a) EP-SCE and (b) CONV observed using GDOES.

but is not observed in the compression test of as-deposited EP-SCE pillar. Figure 25 shows the SEM image of pillars after the compressive deformation. In the Cu pillar of CONV, many slip traces can be observed on the surface of the individual grains. Therefore, the deformations can be assumed to occur as dislocation activation inside grains. On the contrary, sharp shear band crossing through the pillar is observed in the as-deposited EP-SCE pillar, which explains no linear hardening in compressive deformation. SEM images from different sides indicated the shear sliding of the pillar top. Decrease in the cross-sectional area is followed by shear localization responsible for the work softening behavior in micro-compression of as-deposited EP-SCE. This is because the dislocation storage is strikingly limited by grain refinement when the grain size is in sub-micron meter regime.

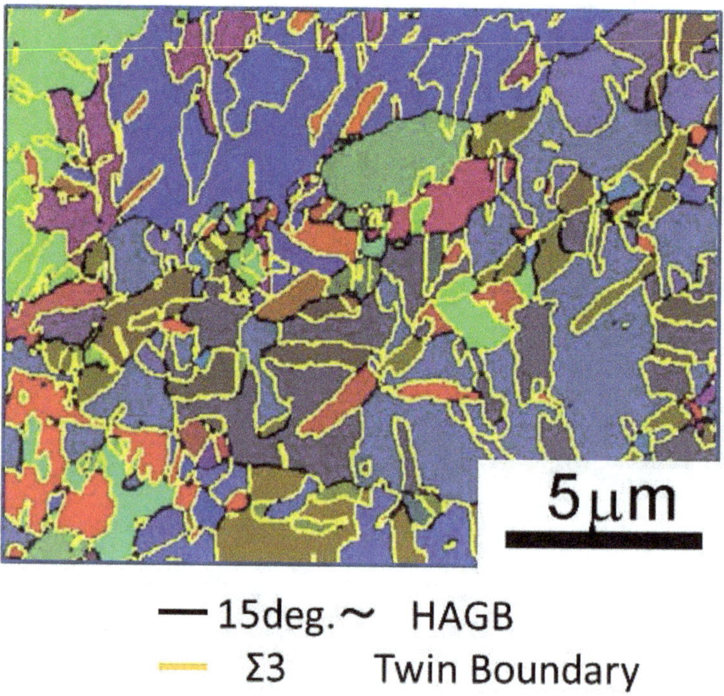

Figure 23. Orientation maps obtained from EP-SCE Cu after two months of storage. Grain boundaries of HAGB and twin boundaries are overlaid in black and yellow lines.

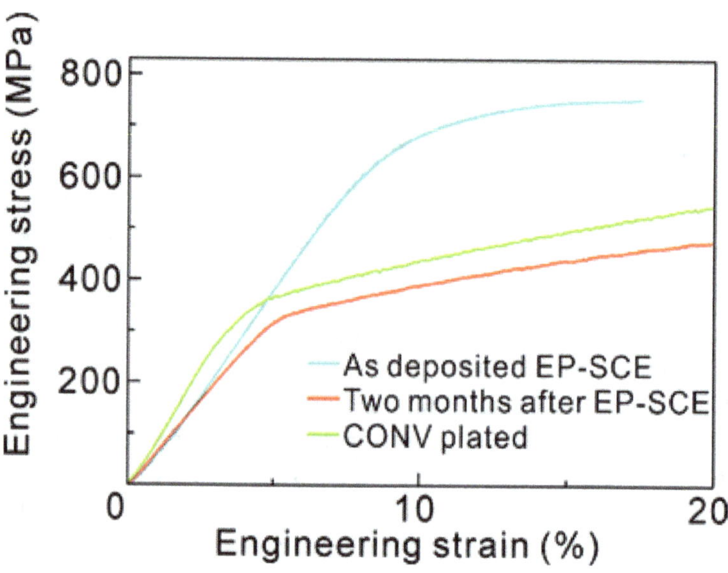

Figure 24. Engineering stress-strain curves in micro-compressions of pillars fabricated from as deposited EP-SCE copper film two months after deposition of EP-SCE copper film and CONV copper film.

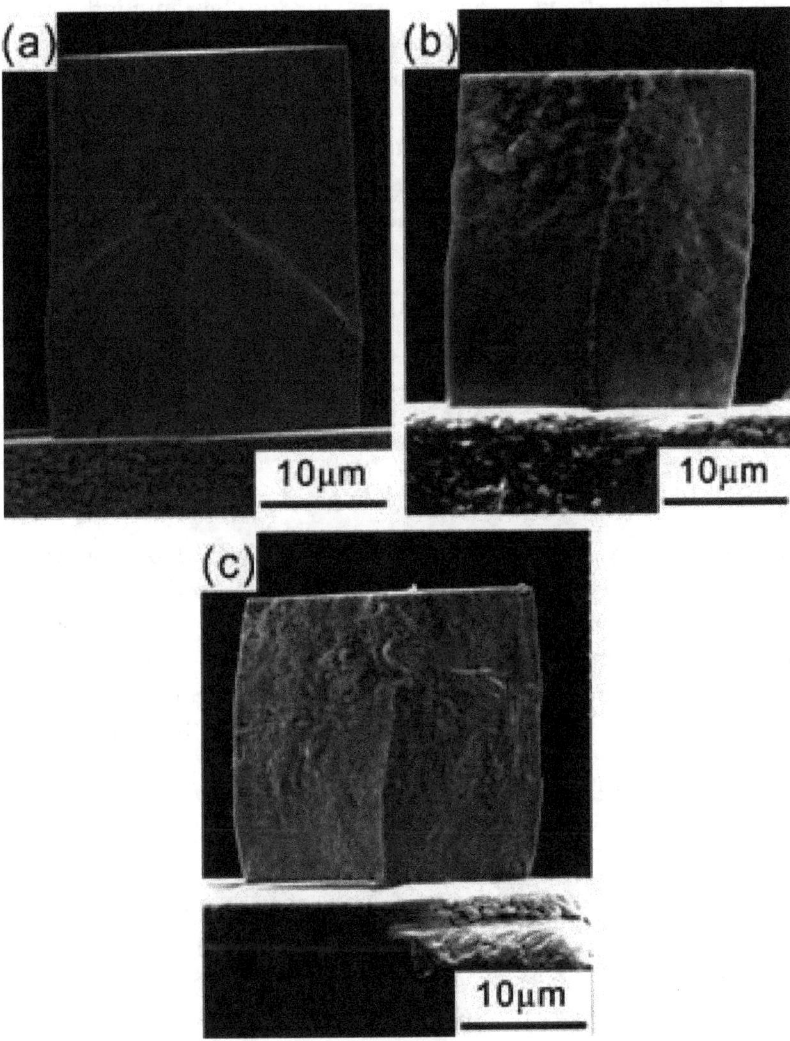

Figure 25. SEM images of post-deformed copper pillar of (a) as-deposited EP-SCE, (b) two months after EP-SCE, and (c) CONV.

Besides limited work hardening, yield stress was increased in the as-deposited EP-SCE pillar compared to self-annealed EP-SCE and CONV pillar. The main reason of strengthening is considered to be grain refinement strengthening. It is widely accepted that grain refinement strengthening is explained as a shortage of dislocation pile-up length due to the decreased distance between grain boundaries. This strengthening contribution was formulated by Hall and Petch and is known as Hall-Petch relationship [67, 68]:

$$\sigma = \sigma_0 + kd^{-\frac{1}{2}}$$

where σ, σ_0, and k are yield strength, friction stress, and the constant for the material, respectively. We calculated the increase in yield strength with decreased grain size from 1.0 to 0.1 μm using the equation. When we refer constant k (MPa·m$^{0.5}$) as 0.14 in previous report on copper [70], the calculated increase in strength was about 300 MPa. This is in good agreement with experimental results of this study. Hence, Cu pillar of as-deposited EP-SCE can be explained only by grain refinement of EP-SCE. Considering the results of GDOES, grain boundary strengthening should be the main factor contributing to the increase of mechanical strength.

4. Conclusions

This chapter describes recent progress in electroplating of micro/nano structures and the evaluation of mechanical properties by micro-testing technique in our group.

Nanocrystalline metals were successfully fabricated using electrodepositon with supercritical CO_2 emulsion. Bouncing micelles desorb hydrogen bubbles and accelerate nucleation by hindering grain growth results in nanocrystalline deposition without defects. Very high strength found in nickel nanocrystals are corresponded to the co-deposited carbon from CO_2. Micro-compression test demonstrated the feasibility for MEMS devices owing to the high strength of nickel. Follow-up micro-testing are also reviewed in this chapter.

Columnar structures of CONV Ni had shown anisotropic mechanical property in micro-cantilever bending. Geometrics of the grains alter the movement of dislocations, resulting in direction-dependent strength of electrodeposited nickel.

Micro-compressions in Cu and Ni revealed size-dependent strength. Materials with S/G ratio smaller than 25 may involve large deviation in mechanical strength. Although in nanocrystalline metals with large S/G ratio tested, increasing strength with decreasing sample size was observed in nickel. This may have corresponded to the CGBS involved in plastic deformation, but further work is needed to explain the detailed mechanisms.

Author details

Takashi Nagoshi[1,2*], Tso-Fu Mark Chang[1] and Masato Sone[1]

*Address all correspondence to: nagoshi-t@aist.go.jp

1 Precision & Intelligence Laboratory, Tokyo Institute of Technology, 4259 Nagatsuta, Midori-ku, Yokohama, Japan

2 National Institute of Advanced Industrial Science and Technology, Tsukuba, Japan

References

[1] Y. Sun, T. Miyasato, J.K. Wigmore, N. Sonoda, Y. Watari, J. Appl. Phys., 2 (1997) 2334.

[2] T. Yoshida, T. Tani, H. Nishimura, K. Akashi, Characterization of a Hybrid Plasma and Its Application to a Chemical Synthesis, J. Appl. Phys., 54 (1983) 640.

[3] M. Gad-el-Hak, The MEMS Handbook; CRC, Taylor & Francis: Boca Raton, Fla., 2006.

[4] J.W. Schultze, A. Bressel, Principles of electrochemical micro- and nano-system technologies, Electrochim. Acta, 47 (2001) 3.

[5] W.L. Tsai, P.C. Hsu, Y. Hwu, C.H. Chen, L.W. Chang, J.H. Je, M.H. Lin, A. Groso, G. Margaritondo, Electrochemistry: Building on bubbles in metal electrodeposition, Nature, 417 (2002) 139.

[6] H. Yoshida, M. Sone, H. Wakabayashi, H. Yan, K. Abe, X.T. Tao, A. Mizushima, S. Ichihara, S. Miyata, New electroplating method of nickel in emulsion of supercritical carbon dioxide and electrolyte solution to enhance uniformity and hardness of plated film, Thin Solid Films, 446 (2004) 194.

[7] H. Yan, M. Sone, N. Sato, S. Ichihara, S. Miyata, The effects of dense carbon dioxide on nickel plating using emulsion of carbon dioxide in electroplating solution, Surf. Coat. Tech., 182 (2004) 329.

[8] H.Yoshida, M.Sone, A.Mizushima, H. An, H.Wakabayashi, K.Abe, X.T.Tao, S.Ichihara, S.Miyata, Application of emulsion of dense carbon dioxide in electroplating solution with nonionic surfactants for nickel electroplating, Surf. Coat. Tech., 173 (2003) 285.

[9] T.F.M Chang, M. Sone, A. Shibata, C. Ishiyama, Y. Higo, Bright nickel deposited by supercritical carbon dioxide emulsion using additive-free Watts bath, Electrochim. Acta, 55 (2010) 6469.

[10] N. Shinoda, T. Shimizu, T.F.M. Chang, A. Shibata, M. Sone, Filling of nanoscale holes with high aspect ratio by Cu electroplating using suspension of supercritical carbon dioxide in electrolyte with Cu particles, Microelectron. Eng., 97C (2012) 126.

[11] T. Shimizu, Y. Ishimoto, T.F.M. Chang, H. Kinashi, T. Nagoshi, T. Sato, M. Sone, Cu wiring into nano-scale holes by electrodeposition in supercritical carbon dioxide emulsified electrolyte with a continuous flow reaction system, J. Supercrit. Fluids, 60 (2014) 60.

[12] A. Shibata, H. Noda, M. Sone, Y. Higo, Microstructural development of an electrodeposited Ni layer, Thin Sol. Film, 518 (2010) 5153.

[13] T. Nagoshi, T.F.M. Chang, T. Sato, M. Sone, Mechanical properties of nickel fabricated by electroplating with supercritical CO2 emulsion evaluated by micro-compression test using non-tapered micro-sized pillar, Microelectron. Eng., 110 (2013) 270.

[14] H. Kinashi, T. Nagoshi, T.F.M. Chang, T. Sato, M. Sone, Mechanical properties of Cu electroplated in supercritical CO2 emulsion evaluated by micro-compression test, Microelectron. Eng., 121 (2014) 83.

[15] M. Tanabe, T.F.M. Chang, T. Nagoshi, S.T. Chung, W.T. Tsai, H. Hosoda, T. Sato, M. Sone, Mechanical properties of Sn electrodeposited in supercritical CO2 emulsions using micro-compression test, Microelectron. Eng., 141 (2015) 219.

[16] J.R. Greer, W.D. Nix, Nanoscale gold pillars strengthened through dislocation starvation, Acta Mater., 53 (2005) 1821.

[17] G. Dehm, T. Wagner, T.J. Balk, E. Arzt, B.J. Inkson, Plasticity and interfacial dislocation mechanisms in epitaxial and polycrystalline Al films constrained by substrates, J. Mater. Sci. Tech., 18 (2002)113.

[18] P.A. Gruber, J. Bohm, F. Onuseit, A. Wanner, R. Spolenak, E. Arzt, Size effects on yield strength and strain hardening for ultra-thin Cu films with and without passivation: A study by synchrotron and bulge test techniques, Acta Mater., 56 (2008) 2318.

[19] M.A. Haque, M.T.A. Saif, A Review of MEMS-Based Microscale and Nanoscale Tensile and Bending Testing, Experimental Mech., 43 (2003) 248.

[20] C.P. Frick, B.G. Clark, S. Orso, A.S. Schneider, E. Arzt, Size effect on strength and strain hardening of small-scale [111] nickel compression pillars, Mater. Sci. Eng. A, 489 (2008) 319.

[21] D.M. Dimiduk, M.D. Uchic, T.A. Parthasarathy, Size-affected single-slip behavior of pure nickel microcrystals, Acta Mater., 53 (2005) 4065.

[22] Z.M. Shan, R. Mishra R, S.A. Syed, O.L. Warren, A.M. Minor. Mechanical annealing and source-limited deformation in submicron diameter Ni crystals. Nature Mater., 7 (2008) 115

[23] J.R. Greer, W.C. Oliver, W.D. Nix. Size dependence of mechanical properties of gold at the micron scale in the absence of strain gradients, Acta Mater., 53 (2005) 1821.

[24] C.M. Byer, B. Li, B. Cao, K.T. Ramesh. Microcompression of single-crystal magnesium, Scripta Mater. 62 (2010) 536.

[25] J.Y. Kim, J.R. Greer. Tensile and compressive behavior of gold and molybdenum single crystals at the nano-scale, Acta Mater., 57 (2009) 5245.

[26] S. Cheng, X.L. Wang, H. Choo, P.K. Liaw. Global melting of Zr57Ti5Ni8Cu20Al10 bulk metallic glass under micro-compression, Appl. Phys. Lett., 91 (2007) 201917.

[27] B.E. Schuster, Q. Wei, M.H. Ervin, S.O. Hruszkewycz, M.K. Miller, T.C. Hufnagel, K.T. Ramesh. Bulk and microscale compressive properties of a Pd-based metallic glass, Scripta Mater., 57 (2007) 517.

[28] Y.H. Lai, C.J. Lee, Y.T. Cheng, H.S. Chou, H.M. Chen, X.H. Du, C.I. Chang, J.C. Huang, S.R. Jian, J.S.C. Jang, T.G. Nieh, Bulk and microscale compressive behavior of a Zr-based metallic grass, Scripta Mater., 58 (2008) 890.

[29] F.F. Wu, Z.F. Zhang, S.X. Mao. Size-dependent shear fracture and global tensile plasticity of metallic glasses. Acta Mater., 57 (2009) 257.

[30] A. Rinaldi, P. Peralta, C. Friesen, K. Sieradzki, Sample size-effects in the yield behavior of nanocrystalline Ni, Acta Mater., 56 (2008) 511.

[31] D. Jang, J. R. Greer, Size-induced weakening and grain boundary-assisted deformation in 60 nm grained Ni nanopillars, Scripta Mater., 64 (2011) 77-80.

[32] F.J. Humphreys, M. Hartherly, Recrystallization and Related Annealing Phenomena, second ed. Elsevier, Amsterdam, 2004.

[33] S. Nakahara, Growth twins and development of polycrystallinity in electrodeposits J. Cryst. Growth 55 (1981) 281.

[34] A. Bastos, S. Zaefferer, D. Raabe, C. Schuh, Characterization of the microstructure and texture of nanostructured electrodeposited NiCo using electron backscatter diffraction (EBSD), Acta Mater., 54 (2006) 2451.

[35] E. Shinada, T. Nagoshi, T.F.M. Chang, M. Sone, Crystallographic study on self-annealing of electroplated copper at room temperature, Mater. Sci. Semicon., 16 (2013) 633.

[36] K. Fujiwara, K. Maeda, N. Usami, G. Sazaki, Y. Nose, K. Nakajima, Formation mechanism of parallel twins related to Si-facetted dendrite growth, Scr. Mater., 57 (2007) 81.

[37] K. Fujiwara, K. Maeda, N. Usami, G. Sazaki, Y. Nose, A. Nomura, T. Shishido, K. Nakajima, In situ observation of Si faceted dendrite growth from low-degree-of-undercooling melts, Acta Mater., 56 (2008) 2663.

[38] K. Fujiwara, K. Maeda, N. Usami, K. Nakajima, Growth Mechanism of Si-Faceted Dendrites, Phys. Rev. Lett., 101 (2008) 055503.

[39] C. Motz, T. Schöberl, R. Pippan, Mechanical properties of micro-sized copper bending beams machined by the focused ion beam technique, Acta Mater., 53 (2005) 4269.

[40] A. Shibata, T. Nagoshi, M. Sone, S. Morito, Y. Higo, Evaluation of the block boundary and sub-block boundary strengths of ferrous lath martensite using a micro-bending test, Mater. Sci. Eng. A, 527 (2010) 7538.

[41] F. Iqbal, J. Ast, M. Göken, K. Durst, In-situ micro-cantilever tests to study fracture properties of NiAl single crystals, Acta Mater., 60 (2012) 1193.

[42] H. Imamura, T. Nagoshi, A. Yoshida, T.F.M. Chang, S. Onaka, M. Sone, Evaluation of anisotropic structure in electrodeposited Ni film using micro-sized cantilever, Microelectron. Eng., 100 (2012) 25.

[43] T.F.M. Chang, M. Sone, Function and mechanism of supercritical carbon dioxide emulsified electrolyte in nickel electroplating reaction, Surface and Coating Technology, 205 (2011) 3890-3899.

[44] S.T. Chung, and W.T. Tsai, Nanocrystalline Ni-C electrodeposits prepared in electrolytes containing supercritical carbon dioxide, J. Electrochim. Soc., 156 (2009) D457.

[45] N. Yamachika, Y. Musha, J. Sasano, K. Senda, M. Kato, Y. Okinaka, T. Osaka, Electrodeposition of amorphous Au-Ni alloy film, Electrochimica Acta, 53, (2008) 4520.

[46] Y. Hori, K. Kikuchi, S. Suzuki, Production of CO and CH4 in electrochemical reduction of CO2 at metal electrodes in aqueous hydrogencarbonate solution, Chemistry Letters, 11 (1985) 1695.

[47] J.W. Cahn, The impurity-drag effect in grain boundary motion, Acta Metall., 10 (1962) 789.

[48] A.J. Detor, C.A. Schuh, Tailoring and patterning the grain size of nanocrystalline alloys, Acta Mater., 55 (2007) 371.

[49] F. Liu, R. Kirchheim, Nano-scale grain growth inhibited by reducing grain boundary energy through solute segregation, J. Cryst. Growth, 264 (2004) 385

[50] L.G. Wang, C.Y. Wang, First-principles calculations of energies of impurities and doping effects at grain boundaries in nickel Mater. Sci. Eng. A, 234 (1997) 521.

[51] R.L. Fleischer, Solution hardening, Acta Metall., 9 (1961) 996.

[52] C.A. Schuh, T.G. Nieh, T. Yamasaki, Hall-Petch Breakdown Manifested in Abrasive. Wear Resistance of Nanocrystalline Nickel, Scripta Mater., 46(2002) 735.

[53] F. Ebrahimi, G.R. Bourne, M.S. Kelly and T.E. Matthews, Mechanical properties of nanocrystalline nickel produced by electrodeposition, Nanostruct. Mater., 11 (1999) 343.

[54] G.D. Hughes, S.D. Smith, C.S. Pande, H.R. Johnson, R.W. Armstrong, Hall-Petch strengthening for the microhardness of twelve nanometer grain diameter electrodeposited nickel, Scripta Metall., 20 (1986) 93.

[55] C.A. Schuh, T.G. Nieh, H. Iwasaki, The effect of solid solution W additions on the mechanical properties of nanocrystalline Ni, Acta Mater., 51 (2003) 431.

[56] D. Tabor, The Hardness of Metals, Oxford University Press, London, 1951.

[57] W.M. Yin, S.H. Whang, R.A. Mirshams, The Fracture Toughness of Thin Sheet Nanocrystalline Pure and Carbon Doped Nickel, Acta Mater., 53 (2005) 383.

[58] Y. Kihara, T. Nagoshi, T.F.M. Chang, H. Hosoda, T. Sato, M. Sone, Tensile behavior of micro-sized specimen fabricated from nanocrystalline nickel film, Microelectron. Eng., 141 (2015) 17.

[59] T. Nagoshi, T.F.M. Chang, T. Sato, M. Sone, Sample size effect of electrodeposited nickel with sub-10 nm grain size, Mater. Lett., 117 (2014) 256.

[60] M.G. Zelin, A. K. Mukherjee, Acta Metall. Mater., 43 (1995) 2359.

[61] M. Stangl, M.Lipták, A.Fletcher, J.Acker, J.Thomas, H.Wendrock, S.Oswald, K.Wetzig, Influence of initial microstructure and impurities on Cu room-temperature recrystallization (self-annealing), Microelectron. Eng., 85 (2008) 534.

[62] C. Lingk and M. E. Gross, Recrystallization kinetics of electroplated Cu in damascene trenches at room temperature, J. Appl. Phys., 84 (1998) 5547.

[63] M. E. Gross, K. Takahashi, C. Lingk, T. Ritzdorf, and K. Gibbons: Advanced Metallization Conf. 1998, ed. G.S. Sandhu, H. Koerner, M. Murakami, Y. Yasuda, N. Kobayashi, Pittsburgh, PA: Materials Research Society, (1999) 51.

[64] D.A. Porter, K.E. Easterling, M.Y. Sherif, Phase Transformations in Metals and Alloys, Third ed., CRC Press, Boca Raton, 2014.

[65] R.W. Armstrong, On size effects in polycrystal plasticity, J. Mech. Phys. Solids, 9 (1961) 196.

[66] M. Mutoh, T. Nagoshi, T.F.M.Chang, T. Sato, M. Sone, Micro-compression test using non-tapered micro-pillar of electrodeposited Cu, Microelectron. Eng., 111 (2013) 118.

[67] E.O. Hall, The Deformation and Ageing of Mild Steel: III Discussion of Results, Proc. Phys. Soc. London B, 64 (1951) 747.

[68] N.J. Petch, The Cleavage Strength of Polycrystals, J. Iron Steel Inst., 174 (1653) 25.

[69] W.L. Chan, M.W. Fu, J. Lu, J.G. Liu, Modeling of grain size effect on micro deformation behavior in micro-forming of pure copper, Mater. Sci. Eng. A, 527 (2010) 6638.

[70] G.D. Hughes, S.D. Smith, C.S. Pande, H.R. Johnson, R.W. Armstrong, Hall-Petch strengthening for the microhardness of twelve nanometer grain diameter electrodeposited nickel, Scripta Metall., 20 (1986) 93.

10

The Coupled Magnetic Field Effects on the Microstructure Evolution and Magnetic Properties of As-Deposited and Post-Annealed Nano-Scaled Co-Based Films

Donggang Li, Qiang Wang, Agnieszka Franczak, Alexandra Levesque and Jean-Paul Chopart

Abstract

Superimposed external magnetic fields during electrodeposition process offers the possibility to tailor microstructure and properties of the obtained films in a very efficient, contactless, and easily controllable way, what is caused by so-called magnetohydrodynamic (MHD) effect. On the other hand, the nonequilibrium state of as-electrodeposited nanocrystalline films provides a strong thermodynamic potential for microstructural transformation. This means that the benefit effect of magneto-electrodeposition on a nanocrystalline film can be completely consumed by a thermal exposure at relatively low temperature. Magnetic field annealing has been confirmed to be useful for tailoring the microstructure of as-deposited nanocrystalline films for their widespread uses.

The particular interest of this book Chapter "Growth of Co-based magnetic thin films by magnetic fields (MF) assisted electrodeposition and heat treatment" is the finding that the microstructure and magnetic properties of nanocrystalline Co-based alloys and oxides like CoX (X= Cu, Ni, NiP, FeO) are improved by imposition of MF during elaboration process or post-annealing process. According to the previous study, the targeted scientific activities pay more attention to develop alloys and oxides in nano-scale using pulsed electrodeposition assisted by high magnetic field (HMF). (Note: Since the instantaneous current density during pulse electrodeposition is higher than that during direct current plating, the microstructure of the nano-scale electrodeposits can be more easily controlled by perturbing the desorption/adsorption processes occurring in the pulse electrodeposition process).

During the experiment, high magnetic field is an in-situ method for the control of electrodeposition process. The obtained material is then annealed or oxidized after elaboration under HMF. Comparative studies are performed concerning the electro-deposition process in a high magnetic field, by changing the magnetic field parameters such as magnetic flux density, direction, gradient. Then, we will investigate the evolution of the microstructure induced by magnetic field, and the control of crystal orientation, crystal size, and its distribution by a HMF. By comparing the microstructure and magnetic properties of the film with and without a HMF, we can find optimum magnetic field parameters for the control of the growth of nanocrystalline Co-based magnetic film. The functionality of materials could then be improved by both processes under HMF: electrodeposition and annealing.

Keywords: coupled magnetic field effects, microstructure evolution, nano-scaled Co-based films

---

1. High magnetic field annealing dependent on the morphology and microstructure of electrodeposited Co/Ni bilayered films

1.1. Introduction

Electrodeposited nanocrystalline magnetic films have received considerable interest in the fields of data storage devices and microelectromechanical systems (MEMS) [1], since the microstructure of the film can be easily controlled on a large scale during the electrochemical processing in comparison with a physical vapor deposition. It is well known that the physical properties of the electrodeposits strongly depend on morphology and microstructure, as reported in recent papers [2-3]. Numerous studies have been devoted to improve surface morphology and microstructure, i.e., reducing the grain size and the roughness, by magneto-hydrodynamic effect (MHD), which is induced by superimposing an external magnetic field [4-6] during the electrodeposition process. However, the nonequilibrium state of as-electro-deposited nanocrystalline layers provides a strong thermodynamic potential for microstructural transformation. This means that the benefit effect of magneto-electrodeposition on a nanocrystalline film can be completely consumed by a thermal exposure at relatively low temperature [7]. Magnetic field annealing has been confirmed to be useful for tailoring the microstructure of as-deposited nanocrystalline films for their widespread uses [8]. Markou et al. [9] obtained a high-degree (001) texture in $L1_0$ CoPt films by annealing Co/Pt bilayers in a perpendicular magnetic field of 1 kOe. The advancement of superconducting technology has opened unique aspects for heat treatment in a high magnetic field. Up to date, few applications of high magnetic field (over 10 T) during annealing process [10] proved it to be a promising method to control the preferential crystallographic orientation of the deposited magnetic film. Though there is no doubt that high magnetic field annealing plays an important role on grain growth, inter-diffusion [11-13], and recrystallization processes, which in turn will influence the microstructural variation of nanocrystalline film, the dependences of grain shape, grain size, roughness, and texture of electrodeposited films on the magnetic field parameters during post-annealing are still unclear. In this study, we investigate the effect of post-high magnetic field annealing up to 12 T on Co/Ni bilayered film obtained by electrodeposition assisted with or without a 0.5 T magnetic field.

1.2. Experimental

Nanocrystalline bilayered films were electrodeposited on ITO (1000 nm thickness)/ glass (1 cm^2) with the structure of ITO glass/Ni/Co from two cells step by step. One cell contained the electrolyte of nickel sulfate 0.6 M, boric acid 0.4 M, and pH was adjusted to 2.7 at 30℃. The electrolyte in the other cell consisted of cobalt sulfate 0.6 M, boric acid 0.4 M, at 50℃, pH = 4.7. The counter electrode was a rectangular Pt plate of $1 \times 2cm$, placed parallel to the working electrode. Chronopotentiometry method with a current density of 10 mA /cm^2 was performed for 6 min in Ni^{2+} cell, and 4 min in Co^{2+} cell, respectively. A water-cooled electromagnet was used to produce a magnetic field of 0.5 T. The magnetic field was superimposed parallel to the working electrode during the electrodeposition processing of Co film. Two series of electro-deposits were denoted, 0 T-B$_{ed}$ and 0.5 T-B$_{ed}$.

The two series of as-deposited films were heat-treated at 673 K for 4 h under a protective atmosphere of high-purity argon in a vertical furnace, which was placed inside a supercon-ducting magnet of up to 12 T, and then cooled down to room temperature. The direction of the high magnetic fields was vertically upward, in perpendicular to the horizontally placed films, while the strengths of the magnetic fields were 0, 6, and 12T, respectively.

The surfacial morphology of the Co/Ni films was observed by using a field emission scanning electronic microscopy (FE-SEM, SUPRA 35) made in Japan with high magnification times. The topography and the roughness were investigated by the AFM (NTEGRA AURA, NT-MDT, Russia). Each sample was measured in areas of $10\mu m \times 10\mu m$ for three times at different positions of the film to obtain the average grain size and roughness of the surface. For the characterization of the films microstructure, Grazing Incidence X-Ray Diffraction (GI-XRD, Ultima IV, Japan) measurements were performed using the Cu $K\alpha$ radiation (40 kV, 40 mA,

$\lambda = 1.54 A$). X-ray photoelectron spectroscopy (XPS) measurement (ESCALAB250, Thermo VG Company, USA) was done to analyze the distribution of Co and Ni along the cross-section of the samples, by which the thickness of Co layer can also be estimated as about 1200 nm.

1.3. Results and discussion

The FE-SEM images in Figure 1 show the surface morphology of annealed Co/Ni films under high magnetic fields of 0, 6, 12 T. By comparing the Figure 1(a1-a3) and Figure 1 (b2-b3), it can be seen that the application of high magnetic field annealing played a similar role on the morphological evolution in both series: 0 T-B$_{ed}$ and 0.5 T-B$_{ed}$, although there is a obvious difference between the two series, which were annealed at the same 6 T magnetic field condition. It is generally considered that the effects of a weak magnetic field (B$_{ed}$) on the morphology of electrodeposits were almost fully consumed by an annealing process for a long time. The films after post-annealed under 0 T magnetic field present a blurred irregular grain shape with a mass of velour-like deposits on the grain boundary. With the increase of the applied magnetic field, not only the grain shape and grain boundary become clear to form polyhedral grains but also the grain size decreases with the disappearance of the velour-like deposits.

Figure 1. FE-SEM images of 0 T-B$_{ed}$ series Co/Ni film annealed under different magnetic fields (a1) 0 T, (a2) 6 T, (a3) 12 T; and 0.5 T-B$_{ed}$ series annealed under (b1) 0 T, (b2) 6 T, (b3) 12 T. [48]

As it is well known, for magnetic and paramagnetic materials, magnetic field will induce a pulling force due to the magnetization [14-15], which force is favored for the separation of grains. In the meantime, an introduction of the magnetization energy during the magnetic field annealing results in the decrease of diffusion activation energy, which will improve the diffusion of the congregated deposits on the grain boundary. This is a possible reason for the grain boundary becoming clear after post-magnetic field annealing. Another most likely reason for the disappearance of velour-like product, which is estimated to be Co-oxides formed at the grain boundaries, is that the application of high magnetic fields during annealing process is favorable for the reduction of oxides, but the mechanism needs further investigation.

On the other hand, we should also take into account the effect of the magnetic field annealing on recrystallization process and grain boundary migration (GBM) in the films. If magnetic field annealing plays a retarded effect on the recrystallization process as reported by Refs. [16-17] or beneficial effects on GBM due to magnetic ordering as reported by Molodov et al [18], the grain size of the film after post-magnetic annealing should be larger than the case of annealing without a magnetic field. However, no such effects of magnetic annealing on the electrode-posited film were observed in our experiments, since we found that the grain size of the film annealed under a high magnetic field seemed smaller than that of the annealed film under a no-field condition.

In order to quantitatively analyze the effects of high magnetic fields on the roughness and the grain size of the bilayered film, AFM top view ($10 \times 10 \mu m^2$) and three-dimensional AFM image of the surface topography of nanocrystalline films annealed at different magnetic flux densities ($B_{annealing}$) are shown in Figure 2. The grain shape of the films is similar for the two series samples annealed under magnetic flux densities of 12 T, therefore we only provide the AFM images of 0.5 T-B_{ed} series here. After no-field annealed, there are some ellipsoidal grains imbedded in the spherical grain matrix as seen in Figure 2 (a). While annealing in a magnetic field of 12 T, most of the grains are spherical, even some grains connect with each other to form a chain-like structure (Figure 2 (c)).

To characterize the grain size and the surface roughness, the average values were calculated using standard software (Table 1). For 0 T-B_{ed} series, the application of 6 T magnetic field during annealing process does not affect the grain size, but leads to a decrease of surface roughness. The grain size of the films goes through minimum under the influence of a 12 T magnetic field annealing. On the other hand, in case of 0.5 T-B_{ed} series, 6 T magnetic field annealing is enough to decrease the average grain size of the film from 140 to 100 nm.

The same magnetic annealing conditions inducing different results between the two series indicate that the influence of a weak magnetic field on the electrodeposition process can be conserved in a certain extent after post-annealing process. As the alternation of grain size during magnetic field annealing is not in a general correlation with that of surface roughness, some researchers [19] also showed that a decrease of grain size was not necessarily corre-sponding to a decrease in the surface roughness.

According to the results in Table 1, AFM investigations show that smaller grains and lower roughness are formed during annealing in the presence of a magnetic field comparing with the case of annealing without magnetic field. This phenomenon originates from an accelerative effect of the high magnetic fields annealing on the inter-diffusion between Co and Ni films. The initial as-deposited Co film on the surface consists of 20-50 nm sized grains, which distributed nonuniform with a large number of agglomerate as shown in Figure 3. For a given set of annealing conditions, the nanocrystalline matrix consumed by grain growth, the mean grain size continues to increase. At the meantime, inter-diffusion behavior happens between the Co and the Ni films, which results in Co atoms diffuse into the Ni matrix to form Co-Ni alloy. The application of high magnetic fields during annealing process improved the inter-diffusion behavior according to the conclusions in Refs. [11] and [20]. The experimental

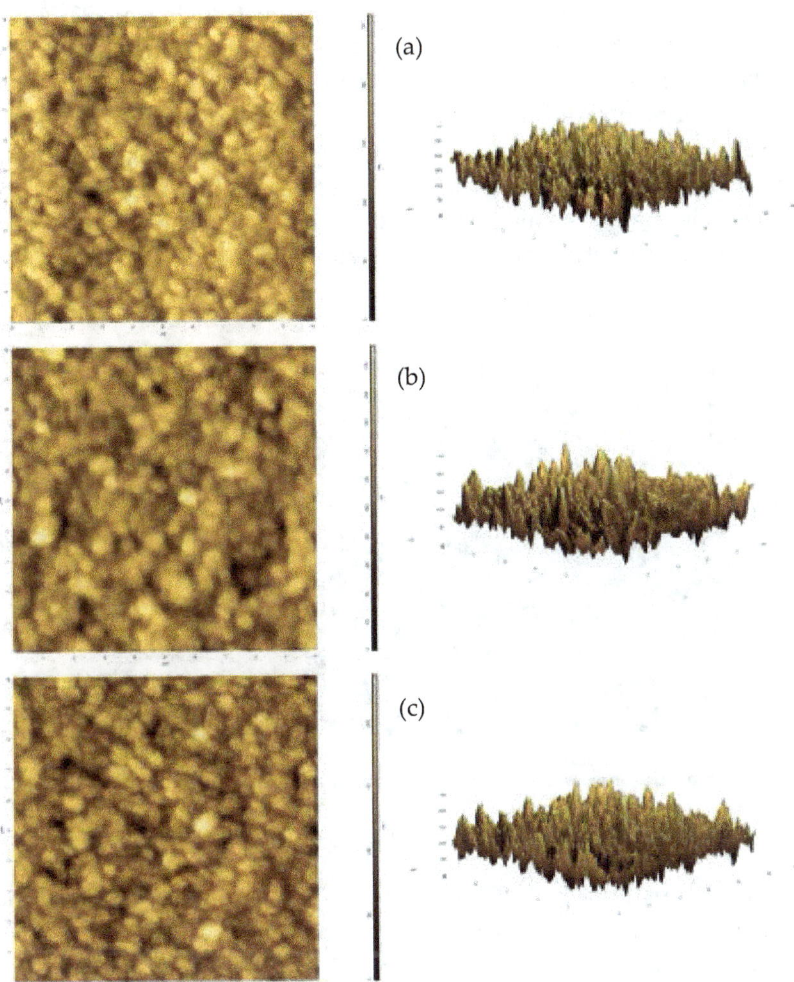

Figure 2. AFM images showing the morphology and roughness of 0.5 T-B$_{ed}$ series Co/Ni films annealed under different magnetic fields (a) 0 T, (b) 6 T, (c) 12 T. [48]

Series	B$_{annealing}$	Grain size (nm)	Roughness (nm)
0 T-B$_{ed}$	0 T	127 ± 5	29 ± 1.9
	6 T	128 ± 3	23 ± 1.8
	12 T	114 ± 3	25 ± 0.9
0.5 T-B$_{ed}$	0 T	140 ± 6	28 ± 1.5
	6 T	100 ± 3	22 ± 1.2
	12 T	104 ± 4	24 ± 1.6

Table 1. Dependence of the average grain size and the average roughness of the two series films on the magnetic flux density of the applied magnetic field annealing.[48]

evidence of this improved inter-diffusion distance by magnetic field from 920 (0 T) to 1330 nm (12 T) was obtained by XPS analysis along the cross-section of the samples. Larger volume fraction of Co-Ni nanocrystalline alloy is formed, smaller is the mean grain size. It is attributed to the crystallographic texture of Co-Ni alloy in $\alpha(fcc)$ phase comparing with $\varepsilon(hcp)$ -Co, which will be discussed in detail in the next section. Another possible explanation for the results in this study is the effect of field annealing on the magnetic domains. The magnetic domains tend to be directional ordering due to magnetization under magnetic annealing, which results in the decrease of the distance between domain walls, as in turn may be responsible for the smaller grain size.

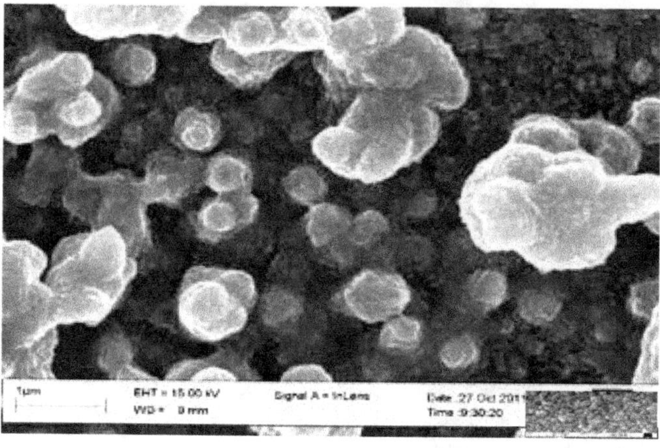

Figure 3. FE-SEM image of the as-deposited Co/Ni film (0 T-B$_{ed}$ series) before annealing.[48]

Figure 4 presents XRD patterns for 0.5 T-B$_{ed}$ series Co/Ni films in the post-magnetic annealing state with magnetic fields of 0, 6, and 12 T (Note: The 0 T-B$_{ed}$ series annealed under a high magnetic field show similar XRD patterns as the 0.5 T-B$_{ed}$ series). It can be seen that all the peaks of hcp-Co and fcc-Ni have been detected, but only one peak (220) of fcc-CoNi have been measured because of the alloying process. The diffraction peak (111) shifts to slightly higher diffraction angles in case of magnetic flux density of 6 T. One reason for the shift is most probably because of the internal stress. Some researchers [9] found the magnetic field annealing can change the internal stress state of films. This behavior can also be rationalized as another reason: more Co atoms are incorporated into the Ni lattice under the moderate magnetic field annealing resulting in a transition from Ni-like microstructural evolution to Co-like microstructural evolution [3].

Since the mean grain size D calculated by Scherrer's formula $D = \dfrac{k\lambda}{\beta \cos\theta}$ is in inverse ratio to β, where β is the full width at half maximum (FWHM) [21]. Here, we list the width and the position of the peaks in Table 2. It can be seen from the data that the width of peaks mostly increases with the increasing magnetic flux density. This also means that the grain size decreased with the magnetic flux density. The change of grain size D is in the same trend as the results obtained from AFM.

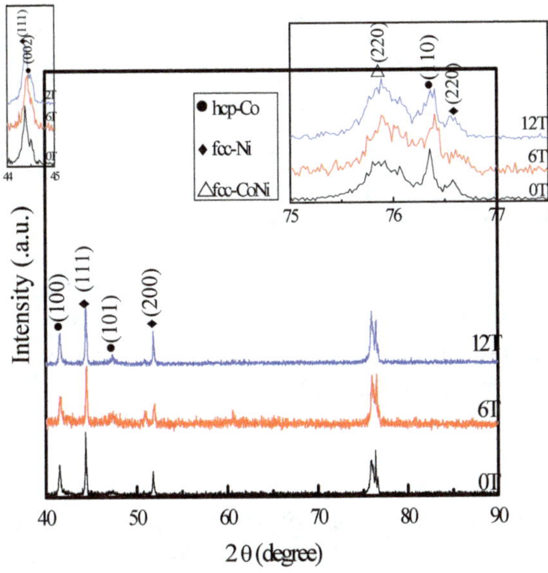

Figure 4. XRD patterns of 0.5 T-B_{ed} series Co/Ni films annealed under magnetic field of $B_{annealing}$ = 0, 6, and 12 T. [48]

There is a general trend in the increase of fcc-CoNi peak (220) and the decreasing of hcp-Co peak (110) with increasing magnetic flux density during annealing process, as can be seen in the amplified zone of Figure 4 at the angular position from $2\theta = 75°$ to $77°$. A higher volume fraction of CoNi alloy in a higher magnetic field was concluded not only from the increase of (200)CoNi compared to the (100)Co shown in Table 3 but also from the increase of diffusion zone according to the XPS composition analysis.

$B_{annealing}$ (T)			0	6	12
FCC-Ni	(111)	2θ (°)	44.44 ± 0.005	44.36 ± 0.011	44.42 ± 0.014
		FWHM	0.180 ± 0.002	0.183 ± 0.005	0.182 ± 0.004
	(200)	2θ (°)	51.84 ± 0.020	51.78 ± 0.012	51.84 ± 0.011
		FWHM	0.157 ± 0.003	0.165 ± 0.006	0.160 ± 0.002
	(220)	2θ (°)	76.64 ± 0.081	76.56 ± 0.014	76.62 ± 0.024
		FWHM	0.148 ± 0.008	0.161 ± 0.004	0.160 ± 0.007
HCP-Co	(100)	2θ (°)	41.54 ± 0.004	41.46 ± 0.013	41.54 ± 0.007
		FWHM	0.184 ± 0.002	0.234 ± 0.003	0.225 ± 0.009
	(110)	2θ (°)	76.40 ± 0.021	76.34 ± 0.022	76.40 ± 0.007
		FWHM	0.222 ± 0.008	0.261 ± 0.010	0.256 ± 0.014
FCC-CoNi	(220)	2θ (°)	75.90 ± 0.081	75.82 ± 0.032	75.90 ± 0.006
		FWHM	0.364 ± 0.010	0.365 ± 0.042	0.360 ± 0.029

Table 2. The width and position of the peaks under different magnetic annealing conditions.[48]

$B_{annealing}$ (T)	fcc-CoNi (220)	hcp-Co (110)	fcc-Ni (220)
0	35.86%	45.52%	18.62%
6	40.53%	41.05%	18.42%
12	42.07%	39.80%	18.13%

Table 3. The dependence of intensity (I) ratio of the three peaks on the magnetic flux density during annealing process. In the table, Value = $I_{fcc\text{-}CoNi}$ /($I_{fcc\text{-}CoNi}$ + $I_{hcp\text{-}Co}$ + $I_{fcc\text{-}Ni}$) × 100%.[48]

The magnetism of the films in a magnetic field is favored for the transformation of $\varepsilon(hcp)$ structure to a more energetically stable $\alpha(fcc)$ structure. The application of high magnetic field annealing induces a $\varepsilon(hcp) \rightarrow \alpha(fcc)$ phase transformation, which may account in the increase of thermal stability of the bilayered films [22].

1.4. Conclusion

Uniform nanocrystalline Co/Ni bilayered films were produced by the electrodeposition assisted with a 0.5 T weak magnetic field. Magnetic fields annealing of up to 12 T were selected in this study in order to investigate the high magnetic field effects on the evolution of morphology and microstructure of the films. A 0.5 T weak magnetic field parallel to the electrode induces modification of the deposits leading to changes in the grain size during magnetic field annealing. Grain shape in the Co/Ni morphology was modified with the grain boundary becoming clearer if a high magnetic field was applied during the annealing process. Atomic force microscopy investigations show that smaller grains and lower roughness are formed during annealing in the presence of a magnetic field comparing to the case of annealing without magnetic field. With increasing magnetic flux density during annealing process, the transformation from $\varepsilon(hcp)$ to $\alpha(fcc)$ can be seen by the reduction of the ε -peak intensity, and the increase in that of α -peak. Overlapping effects of the high magnetic field annealing on diffusion, grain growth, alloying, and phase transformation have been discussed to interpret the experimental results.

Note: The main part, figures, and tables in **Section 1** have been published in Ref. [48].

2. The magnetic properties dependence on the coupled effects of magnetic fields on the microstructure of as-deposited and post-annealed Co/Ni bilayer films

2.1. Introduction

Nano-scaled metallic multilayers received great attention in the recent years because of their unique properties, such as mechanical, optical, electrical, and magnetic, with respect to commonly used alloys. Multilayer materials can be successfully produced by electrodeposition

[23], with thickness varying over a wide range from nanometric films [24-26] to micrometric coatings [27-30]. Moreover, with an accurate control of composition and thickness of each layer within the whole deposit, the internal stress that occurs usually in the microfilms can be reduced to a minimum level.

Electrodeposition is an effective method to produce cobalt-based alloys, but thin films obtained by this way contain some defects, such as rough surface. An application of parallel-to-electrode-surface magnetic field during the process can positively affect the deposits by smoothing their surfaces as well as improving magnetic properties [31-33]. This is caused by magnetohydrodynamic (MHD) and magnetocrystalline anisotropy effects. Magnetohydrodynamic convection induced by Lorentz force is considered as one of the characteristic phenomena in magneto-electrochemical processes [34, 35]. It is assumed that the forced convection resulting from the MHD effect decreases the thickness of the diffusion layer and enhances the concentration gradient, which improves the transport of ions toward the electrode surface. Uhlemann et al. [36] investigated the electrochemical deposition of Co-Cu/Cu multilayer under imposed magnetic field. They have reported that the deposition rate of Cu was nearly two times higher in the process carried out under magnetic field compared to the one without a B-field. Furthermore, the authors observed also an alignment of grains in (111) direction caused by the preferred orientation of Co during the electrodeposition in an external magnetic field. In other works [37-41], the magnetic field effects on nucleation, growth, texture, phase composition, and morphology of the Co layers have been noticed. It has been shown that a magnetic field applied during the deposition process is responsible for two different effects: first, field gradients attract ferromagnetic and paramagnetic materials and repel diamagnetic ones; second, even in uniform fields, magnetic dipoles and therefore the deposits align to the magnetic field direction [42,43]. Some researchers have pointed out that when changing the magnetic field to the orientation of perpendicular-to-electrode-surface, it leads to observe some micro-MHD effects [44]. It would affect roughness and morphology of thin films, while magnetizing force could influence already mentioned magnetic properties of deposits.

It is well known that the annealing process is of a great interest due to the possibilities to decrease an internal stress and to increase the chemical order in the film. In the case of bilayered films, new phases can appear. When the annealing process is carried out under magnetic field, the material properties can be strongly modified. As an example, the Ni-Co system studied by Wang et al. [45] can be given. It was found that the annealing process caused an increase in the wear loss, which was attributed to the decrease of hardness parameter as a result of an increase in the grain size during the high-temperature treatment. Kurant et al. [46] reported a dramatic decrease of magnetic anisotropy upon annealing that was caused by an enhanced diffusion at the metals' interfaces. Harada et al. [47] observed a magnetic field effect during the annealing process on the growth of grains in nanocrystalline Ni deposit produced by the electrodeposition method. Li et al. [48] and Yu et al. [49] determined the effect of vacuum magnetic annealing of Ni and Al co-doped ZnO films on structural and physical properties. Mikelson et al. [50] obtained the aligned solidification structure in Al-Cu and Cd-Zn alloys in a 1.5 T magnetic field. Savitisky et al. [51] and Li et al. [48] found the structure alignment along the direction of magnetic field. All these experimental research works have demonstrated that a magnetic field presents significant influence on the obtained materials.

The present work investigates the Co/Ni bilayered nanocrystalline films produced through the temperature-elevated electrochemical deposition, and modified by annealing carried out also under an external magnetic field. The results indicate an increase of the coercive field of deposited Co/Ni bilayers, when the electrodeposition process was conducted under a magnetic field of 1 T. The annealing processing caused further remarkable increase of the coercive field of as-prepared bilayers that has been preserved under magnetic annealing conditions. The magnetic properties are discussed in terms of samples microstructure. In as-prepared samples the in-plane magnetization was observed, while high temperature treatment, causing micro-structural changes in the film, resulted also in appearance of a small component of magneti-zation oriented perpendicularly to the films' plane that could have been observed by MFM analysis. The induced perpendicular magnetization component in the post-annealed samples was a result of the magnetic field applied in the perpendicular direction to the samples' surface during annealing treatment.

2.2. Experimental methodology

The electrochemical experiments were performed in a conventional three-electrode cell. The working electrode (WE) was a square glass substrate of dimensions of 10×10 mm and 1.1 mm height, covered with ITO (In_2O_3:SnO_2) coating (electric contact layer) and embedded into a cylindrical holder. The counter electrode was made of a platinum plate and the reference electrode was a saturated mercury sulfate electrode (SSE). Electrodeposition process has been carried out in a cylindrical double-wall cell under the conditions listed in Table 4. The electrochemical cell was plunged into the gap of Drusch EAM 20G electromagnet, which delivers a uniform horizontal magnetic field parallel or perpendicular to the electrode surface.

System	Chemical agent	Concentration mol/L	Solution pH	Solution temperature °C	Current density mA/cm²	Deposition time min	Magnetic field amplitude T	
							As-deposited	Post-annealed
Ni/ITO	$NiSO_4.7H_2O$ H_3BO_3	0.6 0.4	2.7	30	-10	1	0	-
Co/Ni/ITO	$CoSO_4.7H_2O$ H_3BO_3	0.6 0.4	4.7	50	-10	1	1	≤12

Table 4. Processing conditions for the Co/Ni bilayer thin films. [56]

The electrodeposition process included two steps, which were related to: firstly - deposition of Ni seed-layer directly on ITO/glass substrate, secondly - deposition of Co layer. The Ni seed-layer deposition was undertaken without an applied magnetic field, while the Co layers deposition was performed under superimposition of a magnetic field with the strength up to 1 T in parallel-to-electrode-surface orientation. Thicknesses of the layers were determined according to the X-Ray photoelectron spectroscopy (XPS) measurements done for the similar, but thicker than considered here, Co/Ni layers reported in Ref. [52]. The difference between these two investigated systems is the deposition time, which in [52] was longer: 6 min deposition of Ni seed layer and 4 min deposition of Co proper layer, respectively. Thus, based on this the estimated thickness of layers considered in this work is approximately 200 nm for Ni and 300-400 nm for Co layer, depending on the magnetic field intensity. Additionally, the roughness of Ni seed-layer determined by AFM was $40 \pm 3nm$, while the roughness of Co secondary layer was $38 \pm 2nm$ when the process was carried out without a superimposed magnetic field. Then, roughness of the Co layer has been slightly improved (decreased) when electrodeposition was carried out under superimposed magnetic field: ~ 35 nm (±3) and 29 nm (±2) under 0.5 and 1 T, respectively. The electrolyte pH was adjusted to proper level by addition of sulfuric acid or sodium hydroxide. Besides the basic salts no other chemical additives, which are usually used to avoid the effect of hydrogen evolution on the films morphology, were added. Boric acid, in addition to its buffer function, is a useful additive compound favoring the formation of low roughness deposits. All of the electrochemical investigations were carried out using the chronopotentiometry method, where a constant current density of -10 mA/cm^2 was applied. The potential of working electrode was controlled by means of a potentiostat-galvanostat VersaSTAT4. The heat treatment process of Co/Ni bilayered films was carried out in a vertical furnace placed into the superconducting magnet with strength up to 12 T and perpendicular magnetic field flux to the samples surface. The films were annealed at 673 K during 4 h under a protective atmosphere of high-purity argon and then cooled down to the room temperature. Bruker D8 Advance and GI-XRD Ultima IV X-ray Diffractometers with CuKα radiation have been employed to obtain XRD patterns using standard θ-2θ geometry. Microstructures of the obtained layers have been investigated by field emission scanning electron microscopy (FE-SEM) Supra 35, Japan. The hysteresis loops have been measured by means of a vector VSM (LakeShore 7410) and an alternating gradient field magnetometer (AGFM, PMC 2900) and the magnetic field has been applied in the in-plane configuration. Magnetic force microscopy (MFM, Bruker Multimode V Nanoscope 8) analysis has been performed in tapping-lift mode with CoCr-coated tips.

2.3. Results and discussion

In order to determine the effects of magnetic fields superimposed during both electrodeposition and annealing processes on the magnetic properties of Co/Ni/ITO bilayer films, room temperature in-plane hysteresis loops were recorded. Figure 5 indicates the magnetic characteristic of as-prepared Co/Ni/ITO films.

It is observed that film deposited under 1 T magnetic field tends to be slightly more difficult to magnetize than film deposited in the absence of an external magnetic field. An increase in coercivity (H_c), from 45 Oe (As-deposited, 0 T) to 78 Oe (As-deposited, 1 T) is noticed, when a magnetic field was superimposed during electrodeposition. The occurred variation in the coercive field cannot be attributed to induced magneto-crystalline anisotropy, as is usually the case in physical deposition processes under the application of a magnetic field. In fact, no significant dependence of the magnetic properties is observed along different directions of the

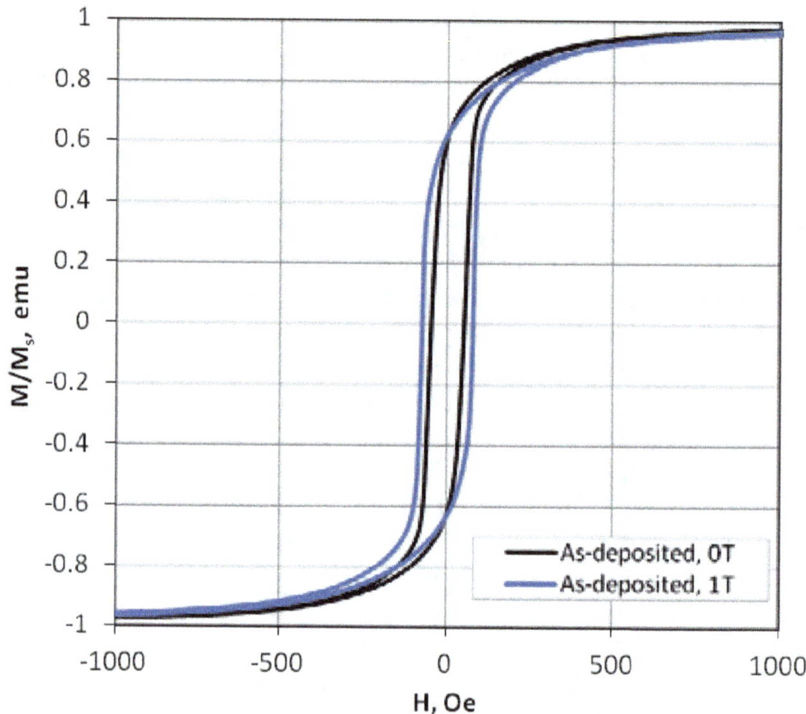

Figure 5. Comparison of the hysteresis loops of Co/Ni/ITO bilayer films electrodeposited without and with a superimposed magnetic field.[56]

applied magnetic field, in the sample plane. Instead, electrochemical deposition under magnetic field affects locally the chemical environment surrounding the magnetic atoms and the mechanisms of grain growth; therefore, the increased value of H_c might be explained in terms of a possible variation of the shape anisotropy of the Co crystals, as reported in [53], or to a slightly larger crystal size when deposition occurs under an applied magnetic field. One of the key issues for the dependence of coercivity on the grain size is the domains around grain boundaries. The rearrangement of the domains costs some energy and therefore contributes to coercivity [54]. The relative volume of the grain boundary domains increases with the increasing of the magnetocrystalline anisotropy, consequently leading to a rise in coercivity.

Both samples were then annealed under magnetic conditions and the hysteresis loops were again recorded. In both cases, as-deposited Co/Ni/ITO films under 0 and 1 T magnetic field, a significant increase of the coercive fields was observed. At the same time, a slower approach to the saturation has been noticed, which is probably due to the increased intermixing of Co and Ni atoms at the interfaces, since their existence has been proven and discussed in Ref. [54]. An increased H_c is mostly associated with the mixed effects of recrystallization process and structural modifications. Further annealing under magnetic conditions of 6 and 12 T brings about only a slight increase of the H_c value (Figure 6), which is most likely to be caused by enhanced increase of Co grains and structural directionality. However, the presence of an interdiffusion zone between Co and Ni layers cannot be neglected as well. In any case, no significant induced magnetocrystalline anisotropy seems to come from the field annealing procedure, as the final coercivity values are mostly independent of the application of a magnetic field during annealing.

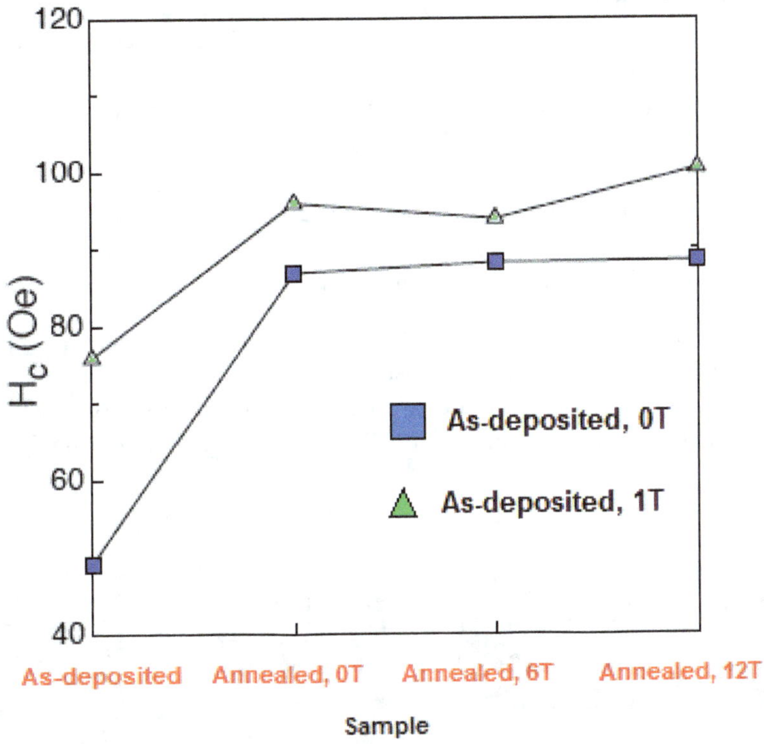

Figure 6. Comparison of the coercivity field of Co/Ni/ITO bilayer films before (as-deposited) and after (post-annealed) annealing treatment.[56]

Extended analysis of the magnetic characteristics of post-annealed Co/Ni/ITO bilayer films have been done by use of magnetic force microscopy (MFM). All samples have been investigated at their in-plane remanence, after a saturating magnetic field has been applied. The MFM tip is magnetized perpendicular to the sample plane; therefore, the MFM is sensitive to the second derivative of the z component of the field generated by the sample magnetization in the half space above the sample surface. The images report in false colors the phase channel in pass 2 (retrace) at a lift scan height of 50 nm. Figure 7 presents MFM images of the Co/Ni/ITO films as-deposited and post-annealed, respectively, on the base of AFM images presented in Figure 2. The magnetic structure consists of bubble-like domains - small, cylindrical magnetic domains that are usually formed when an external magnetic field is applied perpendicularly to the surface of the film. In general, the domains run in random directions in the film plane, i.e., no directionality in the orientation of domains is observed. The domains are displayed as bright and dark areas and have opposite magnetization components perpendicular to the film surface. For the sample deposited under no magnetic field, annealing is responsible for a severe change of the domain structure that is not significantly affected by the application of a magnetic field, in agreement with the coercivity values reported in Figure 6. In particular, the magnetic features size considerably reduces with respect to the as-deposited film, whereas the increased magnetic contrast (as documented by the very different color scales in the top row images in Figure 7) can be attributed to a stronger anisotropy within the individual grains, in turn responsible for the increase of H_c. The Co/Ni films' exposure on the high-temperature treatment changes their microstructure in means of grains' growth. Under thermal conditions, the grains of as-deposited films that were mostly agglomerated into islands begin to grow separately due to the broken symmetry at the interfaces. As a result, ultra-thin magnetic layers tend to be perpendicularly magnetized. Additional contribution might also come from dipolar interaction between the grain islands in the film. In consequence, the observed increasing anisotropy is a result of the single grains growth that boundaries become more clear and visible, and thus, the increased magnetic contrast is observed.

No preferred orientation of the anisotropy is observed even under the application of a 12 T field during annealing. For the sample deposited under the application of a 1 T field on the sample plane, the magnetic features in the as-prepared state are very different with respect to the sample deposited in 0 field, because of the different grain growth mechanisms.

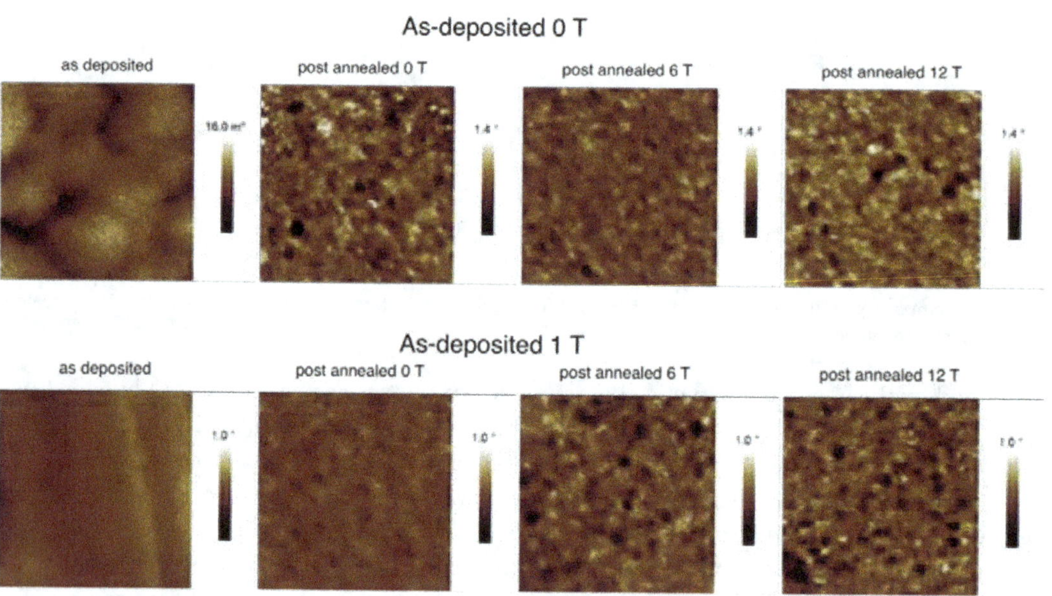

Figure 7. MFM analysis of the Co/Ni/ITO films. The lateral size of each image is 5 μm.[56]

However, after annealing a domain configuration similar to the other sample is obtained, again with no significant dependence on the field intensity applied during annealing. The slightly reduced color scale in the images of the bottom row of Figure 7 with respect to those of the top row account for a slightly lower anisotropy value of the Co grains, which is in agreement with the slightly lower coercivity reported in Figure 6. Moreover, a 2D power spectral density [55] calculated on the MFM images of Figure 7 reports a slight decrease of the magnetic features size after annealing, as evidenced in Figure 8 for two examples, as-prepared sample and annealed at 6 T.

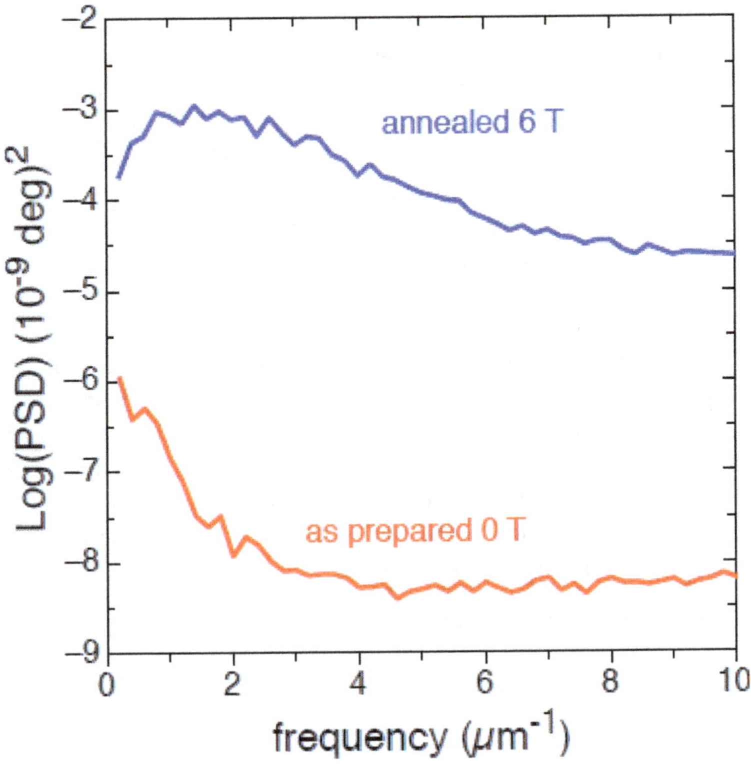

Figure 8. 2D power spectral density of the as-prepared and post-annealed Co/Ni/ITO films.[56]

The appearance of a maximum in the blue curve at higher frequencies (shorter wavelengths) marks the decrease of magnetic features size that can be attributed to a slight increase of the magnetic anisotropy with annealing; a similar behavior has been observed in L_{10} Fe-Pt films [55]. This is in agreement with the measured hysteresis loops, as discussed in Figures 5 and 6. The observed magnetic anisotropy is probably due to the magnetic interactions among the grains: in fact, MFM images (Figure 7) reveal that the magnetic features are larger than the individual grains (AFM images, Figure 2); an increase of the grain size as a consequence of the annealing process increases the dipolar coupling energy among the grains: they are no longer agglomerated into island, but grow separately with larger volume of grain boundaries that results in easier magnetic interaction between them.

In consequence, increasing effective magnetic anisotropy is observed, in agreement with Figure 7, where most grains have preferred orientations in the presence of an external magnetic field. Thus, the magnetic domain structure determines the hysteresis loops of Co/Ni/ITO films, and the domains configuration is affected by the surface morphology, while their reduced size is probably affected by structural defects, such as grain boundaries and impurities.

2.4. Conclusion

The coupled effects of a superimposed magnetic field during both electrodeposition and magnetic annealing treatment in dependence on Co/Ni/ITO microstructure were investigated. The magnetic fields applied in the two processes (growth and annealing) play very different roles. During growth, the modified electrochemical environment around the electrode is responsible for the deposition of Co atoms in islands having a different size and shape anisotropies that result in a higher coercivity. Conversely, it has been shown that the crystallographical structure of as-deposited films is improved by annealing treatment that induces the recrystallization effect. A superimposed magnetic field during annealing promotes diffusion of atoms between Co and Ni layers. The magnetic field annealing has affected the magnetic properties of Co/Ni/ITO films by minimizing the magnetic anisotropy energy of the systems due to the stress relief between grains and directional atomic pair ordering. The direction bonds of dissimilar atoms (Co, Ni) in the films may affect their asymmetric distribution under superimposed magnetic field, when supposing that its strength is sufficiently high. Furthermore, the magnetic field annealing increases the dipolar interactions between metals grains and the redistribution of pair ordering directions, which are perpendicular to the magnetization direction in each domain, as in the case of Co/Ni/ITO films, can be observed. An increased magnetic anisotropy in the investigated films was always obtained after annealing, with coercive fields up to about 100 Oe.

Note: The main part, figures, and tables in **Section 2** have been published in Ref. [56].

3. Influence of high magnetic field annealing on the microstructure of pulse-electrodeposited nanocrystalline Co-Ni-P films

3.1. Introduction

The electrodeposited Co-Ni-P film is a promising magnetic material for micro-electro-mechanical systems (MEMS) [57] and magnetic recording media [58]. The alloy film preparation by pulse electrodeposition method [59] presents the advantage of a nanocrystalline regime and a fine magnetic property, with a low production cost. However, the nonequilibrium state of the as-electrodeposited nanocrystalline layers, in another word, the strong thermodynamic potential for grain growth, may be a limiting factor in the successful applications of this functional film. Recently, the possibility of controlling the grain growth and improving the thermal stability in the as-deposited films by using magnetic field annealing was reported [60 - 62]. Harada et al. [63] annealed the electrodeposited nanocrystalline Ni under a 1.2 MA/ m field and found grain growth could not occur after the early stage. Few applications of high magnetic field during annealing process [64] prove it a promising method to tailor the final microstructure (e.g., surface topography, grain size, crystallographic orientation). In this work, we focus on the microstructure evolution of pulsed-electrodeposited Co-Ni-P film during the magnetic annealing process.

3.2. Experimental

The bath (pH = 3.7) for the electrodeposition of CoNiP films consisted of 0.91 M $NiSO_4$ $6H_2O$, 0.17 M $NiCl_2$ $6H_2O$, 0.11 M $CoSO_4$ $7H_2O$, 0.48 M H_3BO_3, 0.29 M NaH_2PO_2, and 0.0035 M $C_{12}H_{25}NaO_4S$. All electrochemical experiments were performed in a conventional two-electrode cell without agitation. The cathode was Cu plate (10 × 10 × 1 mm) with a seed layer Ni (~60 nm), which was prepared by the physical vacuumed deposition (PVD) method. A nickel plate with area of 2 cm^2 was used as the anode. A unidirectional rectangle pulse current was used, the current was 0.01 A and the duty ratio was 1:5. The pulse-electrodeposition was performed at room temperature for 30 min. After electrodeposited, the samples were heat-treated at 673 K for 4 h under a protective atmosphere of high-purity argon in a vertical furnace, which was placed inside a superconducting magnet. The direction of the external magnetic field was in perpendicular to the surface of the films, and the applied magnetic flux density (B) during the annealing process was up to 9 T. Surface morphology of the Co-Ni-P alloy thin films were observed by field emission scanning electronic microscopy (FE-SEM, SUPRA 35, German). To quantitatively analyze the grain size and the roughness of the films, atomic force microscopy (AFM, NTEGRA AURA, NT-MDT, Russia) was used in the area of 15×15 μm at three different positions to obtain average data. The crystal structure was detected by grazing incidence X-ray diffraction (GI-XRD, Ultima IV, Japan) using Cu $K\alpha_1$ radiation (40 kV, 40 mA, 0.02°/step).

3.3. Results and Discussion

3.3.1. Surface morphology

In Figure 9, surface morphologies of the as-deposited Co-Ni-P film and the annealed films with and without high magnetic field were observed by FE-SEM. It was apparent that the annealing process, whether with or without a magnetic field, induced an obvious evolution in the morphology of the as-electrodeposited film. The as-deposited Co-Ni-P film (Figure 9 (a)) shows a kind of grape-like structure that combined of spherical clusters with different sizes, meanwhile, the surface of the as-deposited film was rough and inhomogeneous. After annealed at 673 K, the small grains in the clusters coalesced together to form bigger grains, with the small grains vanishing at the edges of the bigger grains due to a recrystallization process. However, a 9 T magnetic field annealing favored to form a more uniform surface with smaller grain size and lower roughness, compared with the annealed samples obtained in the absence of magnetic field. The white dots in Figure 9 (b) were phosphorus grains separated out from the alloy, which dots disappeared when a 9 T magnetic field was applied during annealing process. The influencing mechanisms of high magnetic field annealing on the morphology evolution in the CoNiP electrodeposits can be interpreted in terms of the overlapping effects: diffusion and recrystallization. On one side, the introduction of the magnetization energy during the magnetic field annealing decreased the diffusion activation energy [65], resulting in an acceleration effect on diffusion. On the other side, owing to the fact that the retarded effect of a 9 T strong magnetic field on the recrystallization process plays a more important role against its acceleration effect on diffusion process [66-67], the coalescence and growth of small grains were suppressed.

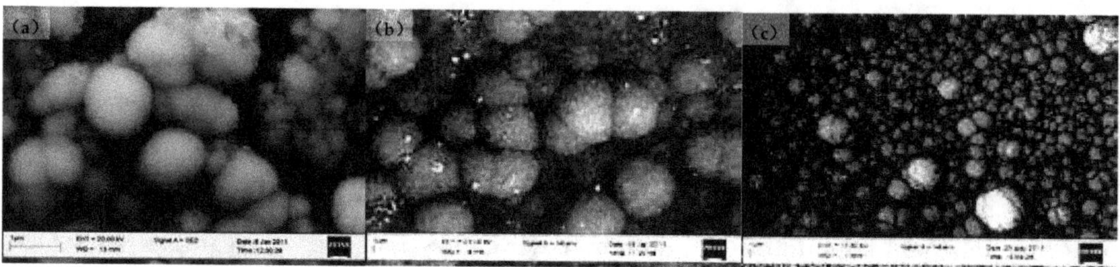

Figure 9. Typical FE-SEM images of as deposited CoNiP thin film (a) and post-annealed in the case of: (b) B = 0 T, (c) B = 9 T.

3.3.2. Average grain size and roughness

In order to quantitatively analyze the roughness and the grain size of the CoNiP films, AFM top view and three-dimensional image of the nanocrystalline films, before annealed, and annealed at different magnetic flux densities (0 T, 9 T) are shown in Figure 10. On the surface of all the films, most grains were in spherical shape with crack-free. Comparing with the as-deposited film, the grain size and roughness in the case of no-field annealing increased dramatically. While the application of 9 T magnetic field during annealing process leads to a decrease in grain size and surface roughness. The average grain size and the average roughness were calculated by standard Nova software attached on the AFM setup. According to the values in Table 5, it was obvious that no significant increase in the grain size of the film after annealed at 9 T field (around 44 nm) in comparison with the as-deposited sample was observed, only a little bigger than that of the as-deposited film (around 32 nm). The evolution of roughness was in the same tendency with that of grain size. Therefore, there is no doubt that a high magnetic field, which plays an important role on grain growth, diffusion, and recrystallization, can be applied to control the grain growth of an as-electrodeposited film during annealing process for improving its thermal stability.

	As deposited	Annealed 0T	Annealed 9T(III)
Grain size (nm)	32.2±3.4	97.53±5.5	44.6±2.6
Roughness (nm)	9.15±0.72	16.35±1.02	8.88±0.56

Table 5. Grain size and roughness of the CoNiP films, before annealed and post-annealed under a 0 T or 9 T magnetic field.

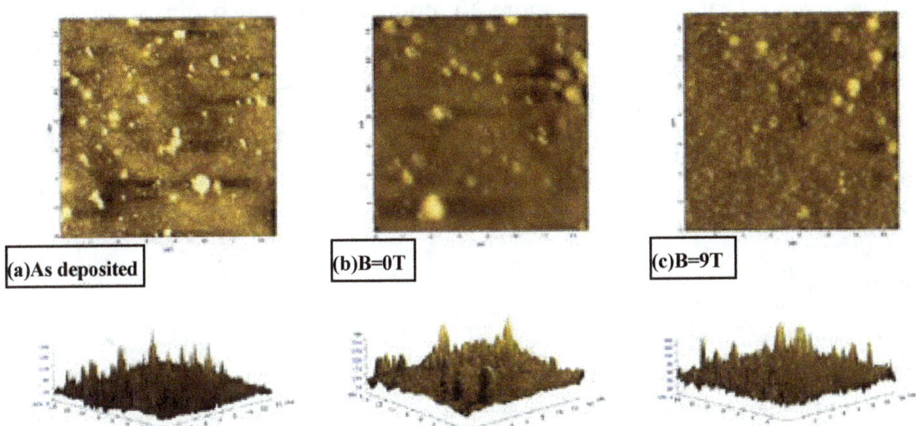

Figure 10. Typical AFM images of CoNiP films, (a) as-deposited and post-annealed under field of 0 T (b) and 9 T (c).

3.3.3. Crystal structure

The XRD patterns of as deposited and post-annealed samples were shown in Figure 11. The magnified diffraction pattern between 42° and 48° of 2θ is shown in the inset of (A).

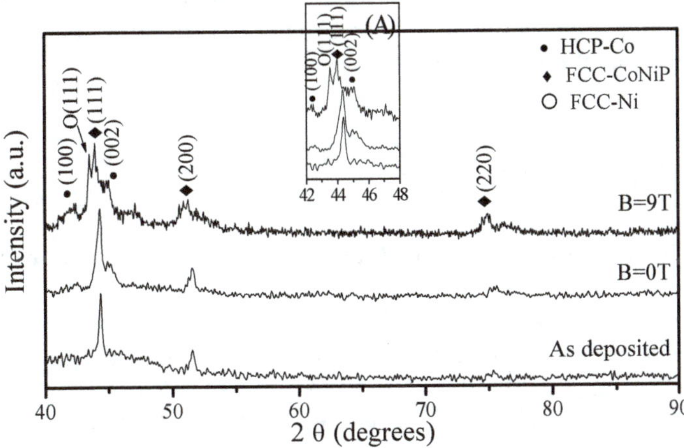

Figure 11. XRD patterns of as-deposited CoNiP film (a), and post-annealed under the magnetic field of : (b) B = 0 T; (c) B = 9 T. A magnified diffraction pattern between 42° and 48° is shown in the inset of (A).

The as-deposited CoNiP film showed a kind of fcc structure which was the α-phase of the Co-based alloy. After annealed at 673 K, an additional hcp-Co (002) peak around 45° appeared, since according to the research of Lew et al. [68] the <001> texture will form after annealed. In the case of annealed under field of 9 T, the (111) diffraction peak of fcc-Ni and the (100) peak of hcp-Co were measured. This means a high magnetic field induces a mixed structure in the film that consists of fcc-Ni, hcp-Co together with fcc-CoNiP alloy. In addition, the diffraction peak (111) of α-phase shifts to lower diffraction angles in case of magnetic flux density of 9 T.

This behavior can be rationalized as more Ni atoms are separated out of the CoNiP solid solution under a high magnetic field, resulting in a transition from Ni-like fcc-microstructural evolution to Co-like fcc-microstructural [69]. The XRD results reveal that during the electro-deposition process, the structure of supersaturated solid solution CoNiP (α-phase), is not stable. After annealing, Ni and Co atoms have a potential to separate out of the alloy, and an accelerating effect of high magnetic field on diffusion modifies this tendency during the annealing process.

3.4. Conclusion

We have studied the influence of high magnetic field annealing on the grain growth and phase structure of pulse electrodeposition CoNiP films. FE-SEM and AFM studies showed that the grains size and the roughness were decreased after annealed under the high magnetic field. X-ray diffraction patterns showed that a mixed phase structure was formed in the annealed film under a 9 T magnetic field.

> **Note 1:** The main part, figures, and tables in **Section 3** have been published on Journal of Iron and Steel Research as: Li DG, Chun WU, Qiang W, Agnieszka F, He J-C. Influence of high magnetic field annealing on microstructure of pulse-electrodeposited nanocrystalline Co-Ni-P films. J Iron Steel Res Int 2012;19:1085-8.

> **Note 2:** The three sections in Chapter one is mainly about the influence of post-annealing under high magnetic fields on the microsturcture of pulse-electrode-posited nanocrystalline Co-based films.

Author details

Donggang Li[1,3]*, Qiang Wang[2]*, Agnieszka Franczak[3], Alexandra Levesque[3] and Jean-Paul Chopart[3]

*Address all correspondence to: lidonggang@smm.neu.edu.cn, wangq@mail.neu.edu.cn

1 School of Materials and Metallurgy, Northeastern University, Shenyang, China

2 Key Laboratory of Electromagnetic Processing of Materials (Ministry of Education), Northeastern University, Shenyang, China

3 LISM, Université de Reims Champagne-Ardenne, France

References

[1] Baghbanan M, Erb U, Palumbo G. In: Mukhopadhyay SM, Seal S, Dahotre N, Agarwal A, Smugeresky JE, Moody N. (eds.) Surface and Interfaces in Nanostructured

Materials and Trends in LIGA, Miniaturization and Nanoscale Materials, The Minerals, Metals and Materials Society, Warrendale, PA, 2004, pp. 307-315.

[2] Cavallotti PL, Vicenzo A, Bestetti M, Franz S. Microelectrodeposition of cobalt and cobalt alloys for magnetic layers, Surf Coat Technol 2003;73:169-70.

[3] Tian L, Xu J, Qiang C. The electrodeposition behaviors and magnetic properties of Ni-Co films, Appl Surf Sci 2011;257:4689-94.

[4] Koza JA, Karnbach F, Uhlemann M, McCord J, Mickel C, Gebert A, Baunack S, Schultz L. Electrocrystallisation of CoFe alloys under the influence of external homogeneous magnetic fields—Properties of deposited thin films, Electrochim Acta 2010;55:819-31.

[5] Georgescu V, Daub M. Magnetic field effects on surface morphology and magnetic properties of Co-Ni-P films prepared by electrodeposition, Surf Sci 2006;60:4195-9.

[6] Krause A, Hamann C, Uhlemann M, Gebert A, Schultz L. Influence of a magnetic field on the morphology of electrodeposited cobalt, Magn Magn Mater 2005;290-291:261-4.

[7] Hibbard G, Erb U, Aus KT, Klement U, Palumbo G. Thermal stability of nanostructured electrodeposits, Mater Sci Forum 2002;386-388:387-96.

[8] Li YB, Lou YF, Zhang LR, Ma B, Bai JM, Wei FL. Effect of magnetic field annealing on microstructure and magnetic properties of FePt films, J Magn Magn Mater 2010 ; 322:3789-91.

[9] Markou A, Panagiotopoulos I, Bakas T. Effects of layering and magnetic annealing on the texture of CoPt films, J Magn Magn Mater 2010 ;322:L61-3.

[10] Li DS, Garmestani H, Yan S-S, Elkawni M, Bacaltchuk MB, Schneider-Muntau HJ, Liu JP, Saha S, Barnard JA. Effects of high magnetic field annealing on texture and magnetic properties of FePd, J Magn Magn Mater 2004;281:272-5.

[11] Liu T, Li D, Wang Q, Wang K, Xu Z, He J. Enhancement of the Kirkendall effect in Cu-Ni diffusion couples induced by high magnetic fields, J Appl Phys 2010;107:103542(1-3).

[12] Li D, Wang Q, Liu T, Li G, He J. Growth of diffusion layers at liquid Al-solid Cu interface under uniform and gradient high magnetic field conditions, Mater Chem Phys 2009;117:504-10.

[13] Li D, Wang Q, Li G, Lv X, Nakajima K, He J. Diffusion layer growth at Zn/Cu interface under uniform and graident high magnetic fields, Mater Sci Eng A 2008;495:244-8.

[14] Coey JMD, Hinds G. Magnetic electrodeposition, J Alloy Compd 2001;326:238-45.

[15] Tang X, Dai J, Zhu X, Song W, Sun Y. Magnetic annealing effects on multiferroic Bi-FeO$_3$/CoFe$_2$O$_4$ bilayered films, J Alloy Compd 2011;511:4748-53.

[16] Martikalnen HO, Lindroos VK. Observations of the effect of magnetic field on the recrystallization in ferrite, Scand J Metall 1981;10(1):3-8.

[17] Xu Y, Ohtsuka Y, wada H. Effects of a strong magnetic field on diffusional transformations in Fe-based alloys, J Mag Soc Jpn 2000;24(4):655-8.

[18] Molodov DA, Bhaumik S, Molodova X, Gottstein G. Annealing behaviour of cold rolled aluminum alloy in a high magnetic field, Scripta Mater 2006;54(12):2161-4.

[19] Hu C-C, Wu C-M. Effects of deposition modes on the microstructure of copper deposits from an acidic sulfate bath, Surf Coat Technol 2003;176:75-83.

[20] Zhao J, Yang P, Zhu F, Cheng C. The effect of high magnetic field on the growth behavior of Sn-3Ag-0.5Cu/Cu IMC layer, Scripta Mater 2006;54(6):1077-80.

[21] Culity BD. Elements of X-ray Diffraction, 2nd edn, Addison Wesley Publishing Company, Inc, London, 1978.

[22] Hibbard GD, Aust KT, Erb U. Thermal stability of electrodeposited nanocrystalline Ni-Co alloys, Mater Sci Eng A 2006;433:195-202.

[23] Yamada A, Houga T, Ueda Y. Magnetism and magnetoresistance of Co/Cu multilayer films produced by pulse control electrodeposition method, J Magn Magn Mater 2002;239:272-5

[24] Herman AM, Mansour M, Badri V, Pinkhasov P, Gonzales C, Fickett F, Calixto ME, Sebastian PJ, Marshall CH, Gillespie TJ. Deposition of smooth Cu(In,Ga)Se2 films from binary multilayers, Thin Solid Films 2000;361-362:74-8.

[25] Toth Kadar E, Peter L, Becsei T, Schwarzacher W. Preparation and magnetoresistance characteristics of electrodeposited Ni-Cu alloys and Ni-Cu/Cu multilayers, J Electrochem Soc 2000;147:3311-18.

[26] Gomez E, Labarta A, Llorente A, Valles E. Electrochemical behaviour and physical properties of Cu/Co multilayers, Electrochim Acta 2003;48:1005-13.

[27] Kirilova I, Ivanov I, Rashkov S. Anodic behaviour of one and two-layer coatings of Zn and Co electrodeposited from single and dual baths, J Appl Electrochem 1998;28:637-43.

[28] Jensen JD, Gabe DR, Wilcox GD. The practical realisation of zinc-iron CMA coatings, Surf Coat Technol 1998;105:240-50.

[29] Kalantary MR, Wilcox GD, Gabe DR. Alternate layers of zinc and nickel electrodeposited to protect steel, Br Corrosion J 1998;33:197-201.

[30] Wang L, Gao Y, Xue Q, Liu H, Xu T. Graded composition and structure in nanocrystalline Ni-Co alloys for decreasing internal stress and improving tribological properties, J Phys D: App Phys 2005;38:1318-24.

[31] Armyanov S. Crystallographic structure and magnetic properties of electrodeposited cobalt and cobalt alloys, Electrochim Acta 2000;45:3323-35.

[32] Cavallotti PL, Vicenzo A, Bestetti M, Franz S. Microelectrodeposition of cobalt and cobalt alloys for magnetic layers, Surf Coat Technol 2003;169:76-80.

[33] Yu Y, Wei G, Hu X, Ge H, Yu Z. The effect of magnetic fields on the electroless deposition of Co-W-P film, Surf Coat Tech 2010;204:2669-76.

[34] Chopart J-P, Olivier A, Merienne E, Amblard J, Aaboubi O. New experimental device for convective mass-transport analysis by electrokinetic-hydrodynamic effect, Electrochem Sol Sta Lett 1998;1:139-41.

[35] Levesque A, Chouchane S, Douglade J, Rehamnia R, Chopart J-P. Effect of natural and magnetic convections on the structure of electrodeposited zinc-nickel alloy, App Surf Sci 2009;255:8048-53.

[36] Uhlemann M, Gebert A, Herrich M, Krause A, Cziraki A, Schultz L. Electrochemical deposition and modification of Cu/Co-Cu multilayer, Electrochim Acta 2003;48:3005-11.

[37] Krause A, Uhlemann M, Gebert A, Schultz L. A study of nucleation, growth, texture and phase formation of electrodeposited cobalt layers and the influence of magnetic fields, Thin Solid Films 2006;515:1694-700.

[38] Krause A, Hamann C, Uhlemann M, Gebert A, Schultz L. Influence of a magnetic field on the morphology of electrodeposited cobalt, J Magn Magn Mat 2005;290:261-4.

[39] Franczak A, Levesque A, Bohr F, Douglade J, Chopart J-P. Structural and morphological modifications of the Co-thin films caused by magnetic field and pH variation, App Surf Sci 2012;258:8683-8.

[40] Li D, Levesque A, Franczak A, Wang Q, He J, Chopart J-P. Evolution of morphology in electrodeposited nanocrystalline Co-Ni films by in-situ high magnetic field application, Talanta 2013;110(15):66-70.

[41] Koza JA, Uhlemann M, Gebert A, Schultz L. Nucleation and growth of the electrodeposited iron layers in the presence of an external magnetic field, Electrochim Acta 2008;53:7972-80.

[42] Li X, Ren Z, Fautrelle Y. Alignment behavior of the primary Al3Ni phase in Al-Ni alloy under a high magnetic field, J Crystal Grwth 2008;310:3488-97.

[43] Krause A, Uhlemann M, Gebert A, Schultz L. The effect of magnetic fields on the electrodeposition of cobalt, Electrochim Acta 2004;49:4127-34.

[44] Lee KH, Yoo J, Ko J, Kim H, Chung H, Chang D. Fabrication of biaxially textured Ni tape for YBCO coated conductor by electrodeposition, Phys C 2002;372:866-8.

[45] Wang L, Gao Y, Xu T, Xue Q. A comparative study on the tribological behavior of nanocrystalline nickel and cobalt coatings correlated with grain size and phase structure, Mat Chem Phys 2006;99:96-103.

[46] Kurant Z, Wawro A, Maziewski A, Maneikis A, Baczewski LT. The influence of annealing on magnetic properties of ultrathin cobalt film, Mol Phys Rep 2004;40:104-7.

[47] Harada K, Tsurekawa S, Watanabe T, Palumbo G. Enhancement of homogeneity of grain boundary microstructure by magnetic annealing of electrodeposited nanocrystalline nickel, Scr Mater 2003;49:367-72.

[48] Li D, Levesque A, Wang Q, Franczak A, Wu C, Chopart J-P, He J. High magnetic field annealing dependent the morphology and microstructure of nanocrystalline Co/Ni bilayered films, CMC 2012;30(3):207-18.

[49] Yu M, Qiu H, Chen X. Effect of vacuum magnetic annealing on the structural and physical properties of the Ni and Al co-doped ZnO films, Thin Solid Films 2010;518:7174-82.

[50] Mikelson AE, Karklin YK. Control of crystallization processes by means of magnetic fields, J Crys Grwth 1981;52:524-9.

[51] Savitisky EM, Torchinova RS, Turanoy SAJ. Effect of crystallization in magnetic field on the structure and magnetic properties of Bi-Mn alloys, J Cryst Grwth 1981;52:519-23.

[52] Zhao Y, Li D, Wang Q, Franczak A, Levesque A, Chopart J-P, He J. The accelerating effect of high magnetic field annealing on the interdiffusion behavior of Co/Ni films, Mater Lett 2013;106:190-2.

[53] An Z, Pan S, Zhang J. Synthesis and Tunable Assembly of Spear-like Nickel Nanocrystallites: From Urchin-like Particles to Prickly Chains, J Phys Chem C 2009;113:1346-51.

[54] Rudolf S. Nanoscale Magnetic Materials and Applications, Springer-Verlag US 2009, pp.275-307.

[55] El Asri T, Raissi M, Vizzini S, Maachi AE, Ameziane EL, D'Avitaya FA, Lazzari JL, Coudreau C, Oughaddou H, Aufray B, Kaddouri A. Inter-diffusion of cobalt and silicon through an ultra thin aluminum oxide layer, Appl Surf Sci 2009;256:2731-4.

[56] Casoli F, Nasi L, Albertini F, Fabbrici S, Bocchi B, Germini F, Luches P, Rota A, Valeri S. Morphology evolution and magnetic properties improvement in FePt epitaxial films by in situ annealing after growth, J Appl Phys 2008;103:043912.

[57] Guan S, Nelson BJ. Pulse-reverse electrodeposited nanograinsized CoNiP thin films and microarrays for MEMS actuators, J Electrochem Soc 2005;152(4):C190-5.

[58] Homma T, Suzuki M, Osaka T. Gradient control of magnetic properties in electroless-deposited CoNiP thin films, J Electrochem Soc 1998;145(1):134-8.

[59] Sun M, Zangari G, Shamsuzzoha M, Metzger RM. Electrodeposition of highly uniform magnetic nanoparticle arrays in ordered alumite, Appl Phys Lett 2001;78:2964-6.

[60] Klement U, Silva MD. Thermal Stability of Electrodeposited Nanocrystalline Ni and Co-Based Materials. Proceedings of Sino-Swedish Structural Materials Symposium 2007.

[61] Li YB, Lou YF, Zhang LR, Ma B, Bai JM, Wei FL. Effect of magnetic field annealing on microstructure and magnetic properties of FePt films, J Magn Magn Mater 2010;322:3789-91.

[62] Shamaila S, Sharif R, Chen JY, Liu HR, Han XF. Magnetic field annealing dependent magnetic properties of $Co_{90}Pt_{10}$ nanowire arrays, J Magn Magn Mater 2009;321:3984-9.

[63] Harada K, Tsurekawa S, Watanabe T, Wantanabe T, Palumbo G. Enhancement of homogeneity of grain boundary microstructure by magnetic annealing of electrodeposited nanocrystalline nickel, Scripta Mater 2003;49:367-72.

[64] Zhang X, Wang D, Zhang S. Effect of high magnetic field annealing on the microstructure and magnetic properties of Co-Fe layered double hydroxide, J Magn Magn Mater 2010;322:3023-7.

[65] Li D, Wang K, Wang Q, Ma X, Wu C, He J. Diffusion interaction between Al and Mg controlled by a high magnetic field, J Appl Phys A 2011;105:969-74.

[66] Molodov DA, Konijnenberg PJ. Grain boundary and grain structure control through application of a high magnetic field, Scripta Mater 2006;54:977-81.

[67] Watanabe T, Tsurekawa S, Zhao X, Zuo L. Grain boundary engineering by magnetic field application, Scripta Mater 2006;54:969-75.

[68] Lew KS, Raja M, Thanikaikarasan S, Kim T, Kim YD, Mahalingam T. Effect of pH and current density in electrodeposited Co-Ni-P alloy thin films, Mater Chem Phys 2008;112:249-53.

[69] Tian LL, Xu JC, Qiang CW. The electrodeposition behaviors and magnetic properties of Ni-Co films, Appl Surf Sci 2011;257:4689-94.

Permissions

All chapters in this book were first published by InTech Open; hereby published with permission under the Creative Commons Attribution License or equivalent. Every chapter published in this book has been scrutinized by our experts. Their significance has been extensively debated. The topics covered herein carry significant findings which will fuel the growth of the discipline. They may even be implemented as practical applications or may be referred to as a beginning point for another development.

The contributors of this book come from diverse backgrounds, making this book a truly international effort. This book will bring forth new frontiers with its revolutionizing research information and detailed analysis of the nascent developments around the world.

We would like to thank all the contributing authors for lending their expertise to make the book truly unique. They have played a crucial role in the development of this book. Without their invaluable contributions this book wouldn't have been possible. They have made vital efforts to compile up to date information on the varied aspects of this subject to make this book a valuable addition to the collection of many professionals and students.

This book was conceptualized with the vision of imparting up-to-date information and advanced data in this field. To ensure the same, a matchless editorial board was set up. Every individual on the board went through rigorous rounds of assessment to prove their worth. After which they invested a large part of their time researching and compiling the most relevant data for our readers.

The editorial board has been involved in producing this book since its inception. They have spent rigorous hours researching and exploring the diverse topics which have resulted in the successful publishing of this book. They have passed on their knowledge of decades through this book. To expedite this challenging task, the publisher supported the team at every step. A small team of assistant editors was also appointed to further simplify the editing procedure and attain best results for the readers.

Apart from the editorial board, the designing team has also invested a significant amount of their time in understanding the subject and creating the most relevant covers. They scrutinized every image to scout for the most suitable representation of the subject and create an appropriate cover for the book.

The publishing team has been an ardent support to the editorial, designing and production team. Their endless efforts to recruit the best for this project, has resulted in the accomplishment of this book. They are a veteran in the field of academics and their pool of knowledge is as vast as their experience in printing. Their expertise and guidance has proved useful at every step. Their uncompromising quality standards have made this book an exceptional effort. Their encouragement from time to time has been an inspiration for everyone.

The publisher and the editorial board hope that this book will prove to be a valuable piece of knowledge for researchers, students, practitioners and scholars across the globe.

List of Contributors

Monika Sharma
Indian Institute of Technology Delhi, New Delhi, India

Bijoy K. Kuanr
Special Centre for Nanoscience, Jawaharlal Nehru University, New Delhi, India

Souad A. M. Al-Bat'hi
Department of Manufacturing and Materials Engineering, Faculty of Engineering, International Islamic University, Malaysia

Piotr Tomassi and Zofia Buczko
Institute of Precision Mechanics, Warsaw, Poland

L. Santos, A. Crespo, P. Barquinha, L. Pereira, R. Martins and E. Fortunato
CENIMAT/I3N, Departamento de Ciências dos Materiais, Faculdade de Ciências e Tecnologia, Universidade Nova de Lisboa and CEMOP/Uninova, Caparica, Portugal

J. P. Neto and P. Baião
CENIMAT/I3N, Departamento de Ciências dos Materiais, Faculdade de Ciências e Tecnologia, Universidade Nova de Lisboa and CEMOP/Uninova, Caparica, Portugal
Champalimaud Centre for the Unknown, Champalimaud Neuroscience Programme, Lisbon, Portugal

Abhijit Ray
School of Solar Energy, Pandit Deendayal Petroleum University, Raisan, Gandhinagar, Gujarat, India

Ashutosh Sharma
Department of Materials Science and Engineering, University of Seoul, Seoul, Korea

Siddhartha Das and Karabi Das
Department of Metallurgical and Materials Engineering, Indian Institute of Technology Kharagpur, Kharagpur, India

Yan-Jia Liou and Wu-Jang Huang
Department of Environmental Engineering and Science, National Ping-Tung University of Science and Technology, Ping-Tung County, Taiwan

Masahiko Matsumiya
Graduate School of Environment and Information Sciences, Yokohama National University, Tokiwadai, Hodogaya-ku, Yokohama, Japan

Tso-Fu Mark Chang and Masato Sone
Precision & Intelligence Laboratory, Tokyo Institute of Technology, 4259 Nagatsuta, Midori-ku, Yokohama, Japan

Takashi Nagoshi
Precision & Intelligence Laboratory, Tokyo Institute of Technology, 4259 Nagatsuta, Midori-ku, Yokohama, Japan
National Institute of Advanced Industrial Science and Technology, Tsukuba, Japan

Donggang Li
School of Materials and Metallurgy, Northeastern University, Shenyang, China
Key Laboratory of Electromagnetic Processing of Materials (Ministry of Education), Northeastern University, Shenyang, China
LISM, Université de Reims Champagne-Ardenne, France

Qiang Wang
Key Laboratory of Electromagnetic Processing of Materials (Ministry of Education), Northeastern University, Shenyang, China

Agnieszka Franczak, Alexandra Levesque and Jean-Paul Chopart
LISM, Université de Reims Champagne-Ardenne, France